TODAY'S HOMESTEAD

HOW TO LIVE AN OLD FASHIONED LIFESTYLE IN THE MODERN WORLD

VOLUME I: THE HOMESTEAD SHELTER

Cover photography by Karl Grant

"Today's Homestead," by Dona Grant. ISBN 978–1-60264–224–9.

Published 2008 by Virtualbookworm.com Publishing Inc., P.O. Box 9949, College Station, TX 77842, US.

Manufactured in the United States of America.

Table of Contents

FOREWORD
Realize the Dream!

You've found that perfect piece of land, and laid out the money necessary to make it your own. Whether it consists of one acre, or more than fifty acres, you can see all the potential in it. Perhaps there is a house on it; a grand old farmhouse with a basement, root cellar, and pantry. Then again, maybe it's just a single wide that needs a few repairs or a little paint. Or, like many, maybe you're starting out with just the raw land. No matter what it is, this is going to become your very own homestead. You've probably dreamed about this day for many years, worked hard and salted money away towards making this dream a reality.

Imagine it, if you will. Here, there will be a small orchard for homegrown fruit to feed the family, with excess to sell. A vegetable garden over there, big enough to produce all your vegetables, and of course you'll need a pretty little herb garden over by the house for seasoning and for scent. A berry patch bursting with juicy fruit for all those pies and that homemade jam. A couple of hives of bees over by the fence, and you'll be in the honey business. Of course, you'll need a pasture for the steer or lamb you'll raise for the freezer, or the milk goat or cow you've always wanted. Then, you will need to have a pen for the pigs, and the chicken coop can go over there behind the workshop. Chickens mean fresh eggs and meat. Bear in mind, you'll probably have to cut down some of those trees, and you'll surely need to plow up that turf. How about drilling a well, and cutting some of your own fence posts? Don't forget all the needed repairs you still have to make to your dwelling. Maybe you haven't even started to build the house yet! Does a sawmill to turn all that timber you have to cut into usable lumber sound like something you should invest in?

Before you know it, the idea of making this sweet dream into reality can seem overwhelming. Don't let your dream turn into a nightmare. Prioritize your projects and keep yourself focused on the most important tasks at hand. Make a list and try to finish one job before moving onto another one. That's a difficult endeavor in itself when there's so much work to be done. Try to keep the ultimate goal firmly rooted in your mind.

Keep your eyes open when looking at your property. Recognize and make the best use of your property's own natural assets. Are there any springs or creeks on the property? Is it forested, or nearly all grassland? Nut trees, wild berry bushes and fruit trees are all wonderful ways to take advantage of nature's free bounty as well as adding variety to your homestead's production. Grassland and open fields are pastures that are just waiting to be grazed.

The running of the home itself is also an important part of homesteading. Can't you already smell and feel the rich lather of your own homemade soaps, delicately scented with your own homegrown herbs from that herb garden? Handmade beeswax candles not only add beauty and a touch of elegance to your home, but help the family budget out as well. These candles would be considered an extravagance if you had to buy them in a specialty shop, craft store, or boutique. You also won't find any store or bakery-bought bread that can come close to the delicate texture and flavor of a homemade loaf. The perfume of baking bread is better than any commercial air freshener to sweeten the air inside the home. Does the idea of shelves lined with jars of home-canned produce, put up at the peak of freshness straight from your own garden, fill you with a sense of security and accomplishment? Can you picture the handmade quilts and rugs scattered throughout your home which speak of thrift, talent, and the pride of doing? How about the wood stove in the corner, offering its warmth on a cold day, the thrift of inexpensive or free fuel, and a security that cannot be bought at any price.

These are just some of the delights to be found within the walls of the homestead dwelling.

There is no certain size, shape or location to the perfect homestead. Your perfect homestead will be unique to yourself and your family in the way it fulfills your needs and desires for living a good life. For every homesteader, those needs will likely be different. Never try to tailor your homestead to someone else's ideal. It is as individual as you are, and you should make every effort to make it truly your own.

The first thing you need to do, preferably before the purchase is made, is to educate yourself on your property's deed restrictions, if any, and your area's zoning restrictions. Some residential areas will not allow you to own livestock, especially one within the city limits. It may still be possible to homestead these areas, depending on the specific restrictions. Vegetable and herb gardens are allowed practically everywhere. Fruit trees and berry bushes can be added to an edible landscaping design, thereby fitting into the tightest of spaces. Rabbits and pullets can be raised many places that disallow larger livestock and can be used for live sale, eggs, meat, and fiber, as well as pelts sold from rabbits. Even honeybees can be raised in such an environment by renting out hives to neighboring farms for increased pollinations of their crops. This takes careful pre-planning, but can be accomplished. Canning and freezing, soap and candle making, quilting and rug making, can all be done on the smallest of homesteads. If you have an inexpensive outlet for purchasing goat's or cow's milk, you can make various cheeses for your family in your own kitchen. The things you can do on the smallest homestead are limited only by your own ability and creativity.

CHAPTER ONE
THE PERFECT HOMESTEAD

IN REALITY THERE IS no perfect homestead other than the one you have envisioned in your own head. Homesteads are not "one size fits all". You will need to find the land and home that comes the closest to your vision of the perfect homestead, and then be willing to put whatever money, time, and energy it takes into making it into that perfect homestead.

Although you can homestead, in some sense of the word, on a mere half acre lot, you want to get the maximum amount of land that your skill level and budget will allow. Know that you will grow in experience and your skill level will rise, however your land doesn't usually get any bigger. It's better to purchase a piece of land slightly larger than needed knowing you will eventually grow into it then to purchase something that you will soon outgrow and be dissatisfied with.

First, however, you have to know what you can afford. It's best to approach homesteading as you would any other major investment. Your homestead will probably be the largest investment, in both work and money, that you will ever make. The potential return on your investment is huge, if not from a financial standpoint, then from a security and personal fulfillment one. The single most important thing you can do to prepare yourself for this journey into homesteading is to prepare a budget. Know what you can afford, and stick with it. Look for property that can sustain the life you desire without requiring excessive amounts of cash to make it usable. An acre of land with a rotting old house that you can buy for next to nothing, but that needs an infusion of tens of thousands of dollars and several weeks of round the clock work just to make it livable, isn't a great buy if you don't have the tens of thousands of dollars and the time required to invest right away. It's an even worse investment if you have to hire out to get the needed repairs done. Always keep your budget, and your personal skill level in mind when shopping for real estate for homesteading.

Certain land features are most desirable for a potential homestead. Natural waterways add value and function to the land. A woodlot, however small, is also a bonus. Even if you only have a few trees around the home itself, they will save you a lot of money in heating and cooling costs.

The lay of the land itself is not such an important issue because even the steepest land can be used to graze livestock such as sheep, goats, and cattle. Terraced gardens can be built onto the side of a steep embankment, thereby making use of land that might otherwise be wasted.

Property that has been cleared, or that is mostly open fields is often more expensive than land that is that partly or mostly wooded, depending upon what part of the country you are in. Even completely forested land can be partly cleared to allow for a home site, garden, and pasture area. The logs can be sold, or milled and used on the homestead. A locust or cedar stand can be utilized for fence posts. Other trees can be harvested for firewood. Just be sure not to over-harvest your woods. If you are looking at a forested lot and plan on clearing every stick and twig off the place, perhaps you should look into acreage that is already cleared. Clearing a large piece of forested land can be a daunting project for anyone. Once the trees are gone you still have the tree stumps and roots to deal with. Even if you have a commercial logging company come in and do the work, the cleanup operation will still be left up to you, and most logging companies leave quite a big mess behind. Before hiring any logging company, go look at a couple of their most recent jobs and talk with the property owners, if possible, to get a clear picture of the company's work.

Then, there are the access issues to consider. If the property is on a road, is it gravel or paved? Who maintains this road, if anyone? Is there a permanent right-of-way easement to the land? Will you have to build a bridge to get past a natural waterway that fronts the property? Consider the cost of installing a driveway. How long will the driveway have to be to reach the home? Will you continue the driveway all the way to the barn? What kind of driveway are you considering? Gravel is the least expensive and requires the least upkeep. Concrete and tar are more permanent surfaces that add real curb appeal to your home. Initial cash outlay as well as occasional repairs for either of these driveway surfaces tends to make them more costly alternatives. Any driveway that is cut up a steep incline will be very costly, both immediately and in the future. The steeper the driveway, the more attention that should be paid to natural water run-off. Culverts will need to be put in and gravel will have to be replaced more often to prevent complete or partial washout during heavy rains. Before having any driveway cut, compare bids from several different contracting companies.

Proper water drainage is another issue that any potential homesteader should consider when looking at properties. In many parts of our country natural waterways, and even swamp land, are considered natural wetlands, and your options for homesteading them are greatly reduced because you cannot drain them legally. If your property is wet and you are able to drain it legally, you may want to consider putting in a farm pond or laying underground drain pipe and draining the land into a nearby creek or pond. Either decision carries a fairly big price tag with it, however the benefits often far outweigh the costs. Be sure when looking at wet land that you consider these costs before committing yourself on the property.

When looking at any potential homestead you will also need to consider the utilities that are available to the property. Is there electricity close by or will you have to pay to have it brought in? That can be a very expensive undertaking. Will you live off-grid, or without power other than what you can produce on your own via generator or solar energy system? Is there access to phone lines? If phone lines are not available, is there cell phone service? Very few of us want to be so isolated that we cannot call 911 if we need to.

The homestead shelter is what your family will call home. Be it a mansion, a modest ranch style, or a single wide, this is the center of daily life on the homestead. As homesteaders we generally want to surround ourselves with the products of our efforts. Nothing says comfort like a rocking chair, draped with a sheepskin of our own making, pulled before the inviting warmth of a real wood fire. There's a cup of herb tea steaming on the table, made with our own home grown herbs, of course, and a plate with a slice of oven-warm, homemade bread with fresh churned butter melting over the golden crust, a dollop of home canned raspberry jam, or perhaps a spoonful of your own sweet, golden honey drizzled over the top. This is a delight to be had in the humblest of homestead dwellings.

When shopping for your homestead shelter, take into considerations your budget, as well as your own personal skill level in maintaining or making any needed repairs to the structures on the land.

Small homesteads can be found at very reasonable prices in rural areas fairly easily.

Don't fail to look into the repossession/foreclosure market in your desired area of the country. Quite often foreclosed homes on land can be had for a fraction of their actual value. These homes are often sadly in need of repairs and remodeling. While the initial cash layout required to purchase a repossessed home will be substantially less, there will be an ongoing investment required in labor, cash, and inconvenience until the work is completed. Most homesteaders find the trade-off well worth it, provided they have the time and money to invest, as well as the skill to do much of the needed work themselves.

Some things to look for when touring potential houses are the basic structure of the dwelling, the sanitary systems of the home, as well as the water availability.

A structurally sound roof and foundation are the two most important aspects to look for when considering the purchase at any existing homestead dwelling. A saggy or leaky roof or a crumbling or cracked foundation can consume your entire annual homesteading budget if you aren't careful. If the repairs to any existing home are so extensive that you are not able to do the work yourself, you should rethink that homestead, unless the price is so good that you can easily afford to have the work done for you by professionals without blowing your budget.

Inspect the home's basement or crawlspace. Is there obvious mold growing? Every house has some mold growth somewhere, but if there is an overgrowth of mold it suggests a drainage or flooding problem. Is there standing water in the basement? If the home is very old, is there mud instead of dirt or gravel in the cellar? Check the floor joists for any signs of either damp or dry rot, or insect damage. These situations could lead to costly repairs down the road. Do the floors sag? This could be merely due to settling over the years, or it could be from serious foundation issues. While any of these conditions can be fixed, the solution could end up costing you considerably.

Does the home's wiring and plumbing meet with local code? Many older homes need to have the wiring and plumbing updated, and if you can do these jobs yourself it's not that big of a problem. If you need to call in an electrician or a plumber it could quickly become a problem.

What type of septic/sewer system or services does the homestead offer, or what is available nearby? If the home has a private septic system installed you can search county health department records to determine the year the septic system was installed as well as the type of system installed. City sewer service is often not available in the more rural areas of the country, and when it is available it may be too costly in the long run to be a viable alternative. Some communities will offer a financial incentive or grant to help you connect your home to any existing sewer system rather than continue with a private septic system. Don't be afraid to ask questions. Often times, grants and incentives aren't voluntarily made available.

If you will need to install a private septic system, get in touch with the county officials to learn what types of septic systems are allowed in your area of interest. Mound systems can be enormously expensive, and are usually not realistic in a homestead budget. Sand filtration systems are considerably less expensive then the mound system, but slightly more expensive, and more efficient, then a leash field system. However, a sand filtration system is only an option where the land has certain qualifications to it. The county officials in your particular area will be able help you decide what is best for your property. You will need to schedule an inspection which usually has an upfront cost. You will then have an opportunity to meet with the engineer who will go over your property with you and will make his recommendations in writing to you. You will then want to take that recommendation to various contractors and get bids on the construction of your system. Never accept the first bid before back-up bids are in. Even if it's Uncle John who offers to do the work for you, get competing bids. With each bid, you are not only considering the cost of installation, but warranty and reliability issues as well.

Water availability is another concern that is of premiere importance to any homesteader. As mentioned earlier, when looking over potential homesteads you should keep an eye out for any well, spring, creek, pond, or other natural waterway. Ponds and creeks would best be utilized to provide water for irrigation and livestock use rather than for human consumption. Water is easily pumped out of a pond or creek and piped into the area it is needed.

Clean potable water is of utmost importance to the homestead shelter. Is the home connected to county or town water? If not, is there county or town water available? What is the cost to hook up to a public water supply? You'll also have to consider the monthly expenses associated with any community water provider. Would it be a wiser choice to drill a well? Find out how deep the neighbors' wells are. What is their water quality? The answers to these questions will help you in making the important decisions necessary in buying a homestead.

Keep in mind that there are small well-drilling rigs that can be purchased by the homesteader for less than the cost of having a well drilled commercially. These little rigs are usually a one or two person operation. Not only can you drill a well for the household water supply, but you can drill a second well for use by livestock or in the gardens at no extra cost other than the well casing and pump assembly. They can be a little difficult to use at first, but experience will help you in recognizing when you've hit water. After you've drilled a couple wells on your own property, you can hire your well-drilling rig out to make a little extra cash as well.

If there is a spring close by, can it be used for drinking water by piping it into the house? Is there an existing well? Potential homebuyers always have the right to subject the home's water supply to county or private water testing before purchase. This may be something you want to consider doing. It isn't terribly expensive, and if the water quality is unknown, as with many private water systems, having the water tested may bring some added peace of mind.

If you are fortunate enough to build a new home on your property, you will be able to build it to suit your individual needs, whatever they may be. Be sure, when situating your home on your lot, that you place it in such a way that it faces south or at least maximize the use of your southern exposure. This can make your home much more comfortable in the winter months.

There are a few structural requirements, regardless of the style of shelter you have. The first is the most important. Your home should be well insulated against the weather extremes of your area. To keep heating and cooling bills low, and to maximize on wood heat efficiency, one should invest in the insulation necessary to bring the home up to current standards for your area. This can be done with insulation batts, rolled insulation, or blown in insulation, depending upon your home's needs.

Weatherizing your home should include replacing or repairing any broken or drafty windows. This might involve purchasing new windows, or it might mean using a few tubes of caulk, or replacing a couple of panes of glass. If all the windows in the home need to be replaced, but replacing all the old windows in your home at once is simply out of the budget question, you can replace one or two at a time, while keeping the rest covered with storm windows, a weatherproof window covering, or clear plastic sheeting. The most frequently used room in the house is naturally where you should begin when replacing windows this way, unless of course, there are broken or cracked windows that demand immediate replacement in other parts of the home. The savings realized in installing a good quality double-paned window will be immediate and will continue saving you money year after year. The difference in comfort in your home will be felt immediately as well.

Doors should be checked for drafts and leaks. Often the doors themselves don't need to be replaced, but the weather stripping surrounding the door can simply be replaced.

Old fireplaces can be a huge source of energy loss in your home. While simply closing the fireplace damper can help somewhat with the heat loss, whenever you use the fireplace you are sending far more heat up the chimney than is actually getting into your home. Investing in a fireplace insert is the ideal solution to the problem. Inserts can be bought new or used, and can mean a huge savings in energy wasted. Before any chimney is used, an annual inspection and cleaning should be performed by an experienced chimneysweep. Any chimney damage must be repaired before using the chimney, for obvious safety reasons.

Another consideration for any homestead dwelling is food storage. You're going to have to have a place to store all those root crops, pumpkins, squash, pears, and apples. You'll need shelves for the bottled and canned goods, and freezer space for the winter's supply of berries and pasture-raised meats. A root cellar works wonderfully well for storing those root crops, or, if you've got the luxury of owning a basement, a small food storage room can be built right into a corner. Shelves can be built in the utility room and a small freezer can often be worked into the kitchen or utility room area if there is no basement or root cellar in existence already. You will want to protect your foods from light, warmth, and humidity. You also don't want any food, other then what is already stored in the freezer, to be exposed to freezing temperatures. Certain foods should not be stored in close proximity to one another, such as apples and onions or potatoes. If you have limited space, don't think you will never have the security of long term food storage. There are many other types of outdoor food storage areas that will discussed in a later chapter that will be a big benefit to those who have no other food storage space available.

CHAPTER TWO
Wood stoves and Firewood Basics

There are many types of alternative heating used in today's homesteads. Some people prefer to use propane as a backup to an electric or oil furnace. Others prefer wood as a back-up to propane heat. In some parts of the country coal is inexpensive and readily available. Some folks like the look and feel of a pellet stove. Pellet stoves are nice because they use waste wood that has been processed into a pellet form. They can also burn cherry pits which are readily available from commercial cherry canneries. The main drawback to a pellet stove is that electricity is needed to operate them. A growing trend among homesteaders in the Midwest is grain, or corn stoves. These stoves operate on the same basis as a pellet stove, and have the same drawbacks.

In this book we will be dealing with wood heat as an alternative to other heat sources because wood as a fuel can be obtained for free or nearly so in many instances, and it is an entirely renewable resource. It also does not require the presence of electricity to enable you to heat your home or cook your meals.

As previously discussed, an old, open fireplace means a lot of lost energy in wasted fuel which translates into wasted time and money. Simply purchasing and installing a new or used fireplace insert into your existing fireplace opening is the most cost effective way of eliminating this waste. An old fireplace, with a chimney in good working condition, can have the fireplace opening bricked up, and a hearth laid, or extended, to receive a new wood stove. The stove pipe can simply be exited through a hole made in the chimney for this purpose. A wood stove can also be placed on a free standing hearth, the stove pipe venting up through the roof, or out through the wall. These last two suggestions are often preferred because they allow you to not only heat your home with the wood stove, but to cook and even bake with it. This can lead to substantial energy savings in the winter months. You are getting twice the value from the same amount of fuel used.

Be sure when purchasing your stove to look for the EPA approval plate. These are stoves that meet or exceed EPA emissions rate standards, and when you are talking about using a wood stove in your home and around your family this is very important. There are several different designs that all meet or exceed EPA standards, some more expensive than others. All of them, however, have been proven to burn less wood, and build up less creosote then older non-EPA approved stoves. Kitchen-style wood cook stoves are pretty and very functional. They are also exempted from the EPA standards, however, they are one of the most expensive types of wood stoves to purchase, and don't fit into all home designs as well as some others styles might.

It is vitally important to have your chimney inspected and cleaned before you install your wood burning appliance. Old chimneys may need to be lined before any stove is connected to them as they may be cracked or broken. Cracks will not only allow smoke into your home, but sparks as well could possibly ignite your walls or roof. These small cracks can also greatly interfere with your chimney's draft.

If you'll be burning a lot of green or soft woods you will probably want to do a mid-season stovepipe inspection as well.

If there is a quarter inch accumulation of creosote on the inside of the stove pipe you should brush the inside of the pipe out well. Some people claim to have luck burning a fast hot fire of soft wood in the stove to simply burn off the creosote, however, by doing so you are risking igniting a raging chimney fire. Taking the steps to properly clean the stove pipe is a safer approach to removing creosote buildup. It takes very little time to do this, and can save you grief in the long run.

Besides the annual cleaning and inspection of your chimney and stovepipe, there are some simple maintenance tips that will keep your wood stove looking good and burning efficiently for a long time to come. Keep your wood stove clean and dust free by using any commercial cast iron stove polish, or you can do as grandma did and use a wad of waxed paper, or the inner side of a piece of pig hide, rubbed across the hot surface. Be careful of burning yourself on the hot stove-top! Every year you should inspect your stove's firebrick lining or iron surfaces for flaws and cracks.

Keep a small metal garbage can with a lid on it nearby for ash removal and disposal. Dispose of ash immediately upon removing it from your stove. Take the ash outdoors and spread it on garden areas or compost piles, or place it in a large metal trash receptacle to collect until spring if desired. Remember that ash removed from a wood stove can retain heat for more than 72 hours, so never place ash in a cardboard or plastic container, and keep all ash away from any combustible materials.

Before installing your wood stove you may want to notify your homeowner's insurance provider of your plans. Some local officials require the acquisition of a permit before installation of any wood burning stove. Purchase and install smoke and carbon monoxide detectors throughout your home, and keep and maintain appropriate fire extinguishers on every floor of your home. It's always a good idea to keep an appropriate fire extinguisher close to the stove for easy access in an emergency.

Never use any accelerant such as gasoline, kerosene, paint thinners, or lighter fluids to start or enrich any fire in your stove. This is a recipe for disaster that should be avoided at all costs. A couple sheets of wadded up newspaper and some thin, dry kindling is all that is needed to light a fire in a wood burning stove. Use only plain black and white newsprint for this, never the glossy, colored store advertisers that are often inserted into your newspaper.

Firewood and kindling

Good quality hardwood that is well seasoned and dry will contribute more to your enjoyment of your wood stove than will any other factor. Hardwoods need about 6 to 9 months to season if kept uncovered outdoors. When firewood is kept under cover, the time can be reduced to 4 to 6 months.

Green wood is not recommended to burn in any wood burning device due to the amount of moisture it contains. Green wood requires more air flow to burn well, and most green wood burns smokier then more seasoned woods. The open stove drafts required to burn green wood encourage heat loss via the chimney or stovepipe. Green wood also uses a great deal of the heat it generates to evaporate the water it contains rather than making that heat available for the purpose of heating. Burning green wood also encourages creosote buildup in your stovepipe and chimney..

Even well seasoned hardwood, that is stored uncovered, will be difficult to burn if it is wet with rain or melted snow. It is best to put a covering of some kind over your woodpile, even if it is just a tarp or a plastic sheet. Bear in mind that any firewood attracts insects such as termites, sow bugs, and carpenter ants, and therefore should be stored some distance away from your home. Wood is often sold, and stored, by the cord. A cord of firewood is equal to that of a 4 x 4 x 8 foot space.

Well-seasoned firewood is generally darker in color and has many cracks and splits. It lights easily, even without kindling added. Pound for pound, every species of wood produces the same amount of heat, But denser hardwoods burn slower and generate more heat than less dense woods. Hardwoods such as hickory, beech, oak, sugar maple, and locust are the best woods to burn. Moderately dense woods such as cherry, birch, black walnut, elm, and ash are a good source of firewood as well. The lighter woods such as box elder, willow, poplar are good to add to your woodpile for an even mixture but are not the best woods to burn as a sole source if it can be avoided. These lighter woods, dried well and split into small kindling sticks are excellent for starting a fire on a cold morning. Red cedar also makes a good source of kindling wood. Most of the true softwoods, or conifers, should be avoided for anything other than kindling, if used at all, because of the high creosote content of the smoke they produce. Never burn scraps of exterior plywood, treated telephone poles, or pressure treated wood in your wood stove because of the toxic materials that are released as these woods burn, as well as the damage they can do to your stove.

If you pay for trash removal, you may want to consider burning all paper items in your daily trash, such as newspapers, cereal boxes, envelopes, prepackaged meal boxes, etc. This is an environmentally and economically friendly act. It is amazing how much this reduces your trash volume, and gives you a little extra something to start your morning fire with. Never burn large amounts of paper or cardboard in your stove at one time. Burning large amounts of paper and/or cardboard can result in a hot, flash fire that could easily ignite any creosote buildup in your chimney or stovepipe. It's better to burn paper or cardboard in small amounts over time or use them when initially starting a new fire.

CHAPTER THREE
Soap Making

NOTHING WILL SMELL NOR feel so sweet to you as will your own home-made soap. You are limited in your creations only by your own imagination. You can infuse your soap with your own herbs, flower blossoms, oils, milk, and honey. Of course, you can add oatmeal, strawberries, cucumbers, seaweed, etc. There are a few simple rules to obey when making soap and if these are followed, your soap-making venture should prove safe, enjoyable, and productive.

Four Simple Soap-making Rules

1. Always wear protective clothing like rubber gloves, apron, dust mask, and protective eye wear when making soap.

2. Store caustic soda/lye in a clearly labeled, airtight container, and store it in a safe place away from the reach of chil-

dren.

3. Soap making should be done in a well ventilated area.

4. Keep a bottle of vinegar handy when making soap, and pour on any caustic soda/lye splashes, then rinse with clear water. Splashes to the eye require immediate medical attention.

Soap is made by mixing an acid with a caustic alkali. You can use vegetable oils, and animal fats as the acid, and sodium hydroxide, (caustic soda/lye) as the alkali. When the lye is diluted with water, and mixed with the acid, a reaction called saponification occurs. After letting the soap cure for several weeks the sodium hydroxide has been completely neutralized.

The saponification stage is easy to recognize because when it occurs, your soap has thickened to a point referred to as 'trace'. To decide if your soap has reached trace, simply spoon some soap from the pot and dribble it back over the surface of the mixture. If this dribble sits on top of the mixture, forming a raised line, you have reached trace. Trace can occur in as little as 5 minutes, or it may take several hours. Fragrances, herbs, colorings, and other fillers are added as the mixture reaches trace and the soap is then poured into a mold, and 'put to bed', meaning it is covered with a blanket or heavy towel for insulation, and it is left to set.

Having the right proportions of ingredients is probably the single most important factor in soap making. It is vital to have a small kitchen scale on hand when making soap. Most quantities listed in the soap recipes in this book are weighed quantities, so when liquids are quoted in ounces, these are weighed ounces rather than fluid ounces. To weigh ingredients on your kitchen scales, place an empty container on the scales and turn the setting back to zero. Then, add your ingredients to the container until the weight reads as specified in the recipe. You can use ounces or grams, but you cannot mix the two in any given recipe.

Trace

There are a few ways that you can reduce the time it takes for your soap to get to trace. Using a stainless steel whisk to whip the mixture will speed up the process. You can also use a hand-held electric mixer, provided you set the mixer on the low setting before putting the beaters in the soap mixture.

If you put your soap mixture in the soap pot in an oven that has been set at the lowest temperature possible, you can also speed up the process to trace, however it takes a much more watchful eye this way as your soap can make trace and you not notice, resulting in your soap solidifying in the pot.

Setting Up

The time it takes for your soap to set up, or solidify, is dependent on a number of factors, such as; room temperature, ingredients used, and their temperature, etc. Your soap, on average, should be hard enough to remove from the mold in twenty four to forty eight hours after pouring. Soaps based on vegetable oils generally take longer to harden than those made with animal fats. You should un-mold your soap as soon as it reaches the consistency of hard cheese, because at this point you will still be able to cut it into bars with some ease.

Curing

Soap must be cured before it is to be used. Simply stack the cut bars of soap in a well-ventilated room and cover the stack with a blanket or large towel to continue the curing process. The residual sodium hydroxide in the soap will be neutralized during this process. If a fine white powder or crust appears on your soap, simply scrape or wash it off after the curing process is complete. This is merely soda ash. You can help prevent the formation of this soda ash by wrapping the soap in plastic wrap immediately after it has been poured into the mold. Turn your soap regularly during the curing process to reduce the possibility of the colors fading. This will ensure even light exposure to the bars.

Equipment

There is no reason to have special pots, pans, or utensils specifically for soap making. Simply wash your soap making equipment well when you are finished with each batch, a dishwasher is ideal, and you can use the same equipment for normal kitchen use as well.

Styrofoam trays, cardboard boxes and soda trays, plastic storage boxes, and plastic microwave containers all make excellent soap molds. You can buy commercial soap molds that are elegant and inexpensive. If you mold your soap in a shallow cardboard tray, you can use cookie cutters to cut out fancy soap shapes. When using molds there are a few things to remember.

Soap Mold Rules

1. Cardboard or wood boxes or trays must be lined with a plastic garbage bag before the soap mixture is poured in. Try to eliminate as many wrinkles as possible from the liner so these creases don't show up on your finished product.

2. Never use aluminum cookie trays or baking sheets as the aluminum will come off on your soap.

3. Always grease plastic or rubber molds with solid vegetable shortening or non-stick cooking spray before pouring in soap mixture.

4. If your rigid plastic molds stick to your soap, put the molded soap in the freezer for a few hours. When you take it out of the freezer, set it on the counter until condensation forms, and the soap should pop right out of the mold.

After your soap is set, you can cut it with ease using a cheese slicer or flat bladed kitchen knife.

Fats and Oils

Soap can be made from practically any plant oil or animal fat, or combinations of the two. Among the animal fats, tallow, (beef fat), makes the hardest bar, lard (pork fat), makes a moderately hard bar, and poultry fat makes a softer bar. In the vegetable oils, almond oil makes for a hard bar, olive oil makes a mild, creamy soap that is good for use on sensitive skin or for babies. Some oils make larger bubbles, others offer a more creamy texture. Some oil/fat combinations are especially good for dry skin, others are better for using on oily skin.

Whatever type of fat or oil you choose to use should be clean and free from impurities. Rancid fat should never be used for soap making. Rancid and dirty fats should be cleaned first by boiling it for a few minutes in a ratio of one part fat to four parts clean water. After it has boiled for approximately ten minutes, remove it from the heat and let it cool. When it has cooled and the fat solidified you can remove the fat in one solid piece. You may need to scrape any left-over impurities or dirt off the bottom of the fat block with a knife. As stated earlier, the oil/fat you choose will also affect the time your soap takes to reach trace as well as the consistency of the finished product.

The amount of sodium hydroxide needed to saponify a particular fat or oil will vary greatly, therefore you cannot randomly substitute one fat or oil for another. As you get more comfortable making your own soap, you will want to try working with various oils and fats. You can find saponification charts on the Internet and in books on advanced soap making that will help you to get the right amount of sodium hydroxide for your chosen fat/oil in your soaps.

Following the soap recipes in this book will help you to learn the basics of soap making, and give you a product you can be proud of.

Coloring Dyes

There are several different things that can be used to color soap. Some of these approved items are spices, such as cinnamon, curry, and turmeric. Wax chips are sold for coloring candles and soaps, and even crayons can be used with some success. Most of these items are melted into the soap mixture after it has reached traced. If you try to add them at the beginning of the process, saponification can change the color to something less then you were hoping for. Even adding them at the trace stage can fade or change the color somewhat over time. To use any form of wax colorant, merely heat a half cup or so of traced soap to about 150 degrees F, add the wax chips or crayon, stir to dissolve, and add this mixture back into the main mixture in your soap pot stirring rapidly.

Scenting Oils

Before you decide on the scent you want to add to your soap, you need to decide the purpose of the soap. Most men don't want to use a scrubby-cleansing soap that smells like roses, and you'd probably not want to use a highly spiced soap on an infant's sensitive skin.

There are two types of scenting oils; fragrance oils and essential oils. Essential oils are made by distilling the oil out of the plant it naturally occurs in. A fragrance oil is generally man made by steeping various chemicals in alcohol. Essential oils are the oils generally used in soap making because most fragrance oils come with warnings about using them on skin, although they are generally accepted as safe for home use. They have also been known to seize soap, or create a sticky mess that doesn't saponify correctly. Although essential oils are much more expensive and difficult to find than fragrance oils, they also have better scent retention. Both essential and fragrance oils are often sold under the name of "perfume" oil, so it's important to ask questions to be sure of which you are actually buying.

Be aware that both essential oils and fragrance oils can cause your soap to curdle and, for this reason, it is important to include them in your soap directly after trace. If your soap does begin to curdle or seize, beat it rapidly with a whisk or electric mixer, and pour immediately into your mold.

Re-batching

Re-batching is a process of recycling or reprocessing your soap. Re-batching your soap will greatly reduce the distortion and fading of colorings and fragrances in your soap. Bolder colors and stronger fragrances are achieved when soap is re-batched, and fillers such as herbs and flower petals can be added with a reduced chance of discoloration. By re-batching you can salvage a failed batch of soap rather the throwing it away. Slivers and trimmings of leftover soap can be recycled with the same process, thereby eliminating waste. Soaps made from animal fats make a better product when re-batched then do soaps made from vegetable oils. If you want to re-batch soap made from vegetable oils, let the initial batch of soap cure for at least three weeks before re-batching.

The steps to re-batching are simple:

Using a fine cheese grater, or food processor; grate or grind the soap as fine as possible. Put these gratings into a stainless steel or enamel-lined pan. Mix in a small amount of water, until you have a thin paste. Twelve ounces of distilled water for every pound of soap is a good guide. Set the pot of soap/water mixture over medium-low heat until the mixture melts or liquefies completely. This process can take up to an hour. You can cover the pan lightly if you prefer, but remember to stir the mixture occasionally. If you've added too much water, you can simply allow it to boil off provided you keep the temperature below 230 degrees. If you didn't add enough water the soap will more or less congeal into a sticky ball. Simply mix in more water until the desired consistency is achieved.

When the soap liquefies, remove it from the heat and let it cool. After it has cooled but before it sets up, you can add your fragrance oils, colorants, fillers, etc. If your soap does set up before you get everything mixed into it that you wanted in it, simply return the mixture to the stove and heat until it is liquefied again. After it has been mixed up the way you want it, you can pour it into the molds and let it cool completely. Leave this soap to set as usual.

SOAP RECIPES

Basic Hand Soap

An easy cold process soap for beginners

½ oz (14g) sodium hydroxide (caustic soda/lye)
¼ cup (57g) cold distilled water
½ cup (113g) lukewarm animal fat
1 (15g) Tbsp lemon juice

In a plastic container, gently stir sodium hydroxide into water with a wooden spoon. Slowly add lukewarm fat. Continue stirring until slightly thickened. Add lemon juice, stirring to mix thoroughly. Pour mixture into plastic molds. Cover with plastic wrap and let set for 24 hours. Remove soap from molds and allow it to dry for 21 days. Makes 1 or 2 medium-sized bars.

Goat's Milk Castile Soap

An excellent moisturizing soap made with the heat processed method.

30oz (849g) olive oil
2 oz (57g) beeswax
6 oz (170g) fresh goat's milk
6 oz (170g) distilled water
4 oz (113g) sodium hydroxide (caustic soda/lye)
2 tsp (10g) bay essential oil

1 tsp (5g) pine essential oil
1 tsp (5g) rosewood essential oil

Grease a shallow square or rectangular mold. Place the olive oil and beeswax in a stainless steel or enameled pot over low heat. Warm the goat's milk to room temperature. Pour the milk and water into a heavy glass or plastic bowl. Wearing rubber gloves and eye protection, add the sodium hydroxide to the liquid and stir until dissolved. When the oils have melted together, remove them from the heat. Place one candy thermometer in the oils and another candy thermometer in the caustic solution. When both thermometers reach approximately 130 degrees, carefully pour the caustic solution into the oil. Stir occasionally until the mixture reaches trace. This should take about 20 minutes. When trace has been reached, add the essential oils and stir well. Pour immediately into the mold and cover with a heavy towel or blanket. Leave to set for 24 hours or until the soap reaches a solid consistency. Wearing rubber gloves, remove the soap from the mold and cut into bars. Cover the soap and leave it to cure for four weeks before using.

Honey and Oatmeal Soap

A soothing soap for irritated skin.

15 oz (425g) vegetable shortening
12 oz (340g) olive oil
10 oz (283g) coconut oil
2 oz (57g) beeswax
11 oz (312g) distilled water
5 oz (142g) sodium hydroxide (caustic soda/lye)
2 oz (57g) oatmeal (not instant)
2 Tbsp (30g) honey
2 Tbsp (30g) orange essential oil

Grease a shallow square or rectangular mold. Place the shortening, olive oil, coconut oil and beeswax in a stainless steel or enamel pot over low heat. Pour the water into a heavy glass or plastic bowl. Wearing rubber gloves and eye protection, add the sodium hydroxide to the water and stir until dissolved. When the oils have melted, remove them from the heat.

Place one candy thermometer in the oils, and another in the caustic solution. When both thermometers reach approximately 130 degrees carefully pour the caustic solution into the oils. Stir occasionally until the mixture reaches trace. This should take about 40 minutes.

Sprinkle the oatmeal into the mix and stir well. Add the honey and essential oil.

Stir well. Pour immediately into the mold and cover with a towel or blanket. Leave to set for 24 hours or until the soap reaches a solid consistency. Wearing rubber gloves, turn the soap out of the mold and cut into desired shape bars. Cover the soap again, and leave to cure for four weeks before using.

You can substitute finely ground and dried orange peel for some of the oatmeal in the above recipe for more orange essence.

A Basic Granulated Laundry soap

2 ½ quarts distilled water
2 quarts animal fat, strained and melted
1 lb sodium hydroxide (caustic soda/lye)
3 Tbsp borax

Mix water, sodium hydroxide, and borax. Slowly add strained grease. Remove from heat and leave in pot. Stir often during the first day. Allow two weeks to cure, stirring occasionally.

CHAPTER FOUR
CANDLE MAKING

CANDLES HAVE BEEN USED for centuries for light, security, and comfort. In this age of the stark glare of electricity a gently flickering candle still evokes a sense of domestic bliss. Now, make that candle one of your own creation and you have the ultimate in homestead security and comfort. Candle making is not hard and is very rewarding. Homemade candles make wonderful gifts for friends, and family gift baskets too.

Safety tips

When heating and melting wax, treat it like cooking oil. It is highly flammable and has the potential to cause serious burns. Keep your wax under 212 degrees. Remember, wax will not boil so heating it higher then this is neither necessary nor advisable. The wax will only continue to get hotter and hotter, vaporizing and creating a potential fire hazard. Of course, never leave your melting wax unattended. Overheated wax gives off an acrid, smoky smell. Should this happen, turn off the heat and allow the wax to cool. If you find yourself with a wax fire from overheated wax, turn off the heat and cover the wax pot with a lid to smother the flames. Do not try to remove the wax from the heat as serious burns could result. Never try to douse the fire with water as this will only spread the flames.

Never pour your excess melted wax down the drain because it will set up in your pipes and create a solid blockage. Instead, pour the excess melted wax into an old baking tray, and cut it into squares before it hardens. You can then save it for future use.

If you spill melted wax on carpets or items of clothing, simply scrape off the excess wax and remove the remaining wax by placing a paper towel over the stained area and pressing with a hot iron. The wax will be transferred to the towel. If you should spill melted wax on metal or plastic objects simply place the item in the freezer for an hour or so to make the wax brittle. It then becomes easy to crumble or pick the hardened wax off the item. You can also dip the item into boiling water, which will melt the wax and float it to the surface.

Essential Candle-making Equipment

Candy or wax thermometer
Wax melter, old crock pot, or old fashioned stove top coffee pot
Wick rod or piece of wire coat hanger

Candle waxes

Beeswax has been a favorite candle wax for centuries, due in part to its naturally sweet fragrance and creamy golden color. You can also find bleached beeswax for sale, which has an appealing creamy white color, but less natural fragrance then unbleached beeswax has. Beeswax burns very slowly, and does not shrink as it hardens. However, because beeswax is soft and sticky, it doesn't like to release from molds easily.

Paraffin is a white, translucent wax obtained from the petroleum industry. It is a hard wax that is suitable for many candle-making uses. It releases well from molds and is odorless, although it tends to burn faster than beeswax.

You can buy sheets of beeswax foundation to roll into pretty honeycomb tapers, however, these thin sheets or foundation sheets, are expensive to purchase and are hard on your bees to produce yourself.

Candle scenting agents

Nothing sets the mood of a room more completely than the sweet aroma of a scented candle. Candles can be scented with fragrance oils, essential oils, or herbs and spices.

Fragrance oils are synthetically made scents as opposed to the naturally distilled essential oils. Fragrance oils are less expensive than essential oils and come in a wider variety of scents and scent combinations.

Although essential oils are a costlier alternative to fragrance oils, they are also all natural plant extracts and have an important part in aromatherapy candles.

You can also use finely ground herbs and spices to add fragrance, texture and color to your candles. Finely ground cinnamon, nutmeg, frankincense, and myrrh can all be used to add subtle touches of scent to candles. Only small amounts of finely ground herbs and spices are added because larger amounts will cause your candles to sputter and can affect how your candles burn.

Candle colorants

Candle colorant blocks and commercial candle dyes can be purchased inexpensively at most craft stores, and are probably the best way to add color to your candles. Spices and herbs can add that extra splash of color as well as imparting their own fragrances to your candles. Crayons are not an ideal choice for coloring candles because the melted crayon pigment tends to sink to the bottom of the candle or else is attracted to the wick, making your candle more prone to sputtering. However, you can experiment with various crayons and see if you can find a formula that works for your candles.

Many people enjoy the earthy hue and rich scent of natural beeswax and chose not to add any colorants or scenting agents to candles that contain beeswax.

Mason Jar candles

These candles are fun and easy to make, and they fit into the homestead lifestyle perfectly. They also make great gifts for friends and family.

Place an ample-sized piece of cardboard down on your work surface for protection against spillage. Chop paraffin wax up into small blocks and put them in your wax melter or crock pot. Turn crock pot on high, or melt wax over medium heat, and let the wax melt completely. Do not let wax over-heat, and never leave wax unattended. Once the wax is melted, turn the heat off.

Add your candle fragrance now, a few drops at a time until desired scent is achieved. If using wax color chips, melt separately and add melted chips or liquid color agent to the candle wax at this time. Keep in mind that your candles will be lighter once the wax hardens than it appears in the melted state.

Wear a rubber glove on the hand that will steady the jar to protect your hand from the heat, and using a measuring cup, dip out the wax and fill your mason jars. Save any leftover wax for topping off your finished candles.

Insert your wicks into the jars of wax, either using wick rods or wicks tied onto small lengths of coat hangers that are suspended across the mouth of the jar. You can also put aluminum foil across the top of your jar and poke the wick up through the center of the foil allowing the foil to hold the wick in place.

As your candles harden they will form a well around the wick. After the well forms, simply pierce the wax around the wick with a longer length of coat hanger wire, piercing all the way to the bottom of the container. Bring the leftover wax back up to the melting temperature and fill the well with this. Repeat this process until a well no longer forms. Allow the candles to harden completely. Trim the wicks off even with the top of the jar or ½ to ¾ inches high.

Molded Ice Candles

Ice candles are fun to make and delightful to watch as the lacy patterns slowly burn. Remember when making ice candles that you need to use crushed ice, as larger ice chunks create gaping holes in the wax that can render the candles unstable. Clean, smooth-sided tin cans with only one end removed make good molds for these candles.

Place paraffin wax in wax melter or crock pot. Turn crock pot on high, or put wax melter over medium heat, and melt wax being careful not to overheat wax. Do not leave your wax unattended. Melt colorant chips separately.

You will need to prime your wicks to prevent any water from entering them. To prime a wick simply dip it in melted wax to coat. Let the wick harden or dry completely and dip again. Put wicks aside until needed.

When the wax and colorant have both melted, mix the two together. Keep warm over low heat. Wick your molds using wick hangers or small lengths of coat hanger to hold your wicks suspended across the top of the molds. Spoon crushed ice into the molds bringing it just to the top edge of the mold. Pour the melted wax over the ice until the molds are filled.

When the wax has hardened completely, remove the candles from the molds over a sink and all the melted ice will drain away. If using cans for molds you may have to remove the bottom of the cans and push the candles out. You will now have lacy, unique candles with holes where the ice once was. Allow the candles to dry completely before you light them.

Hand-dipped Tapers

Hand dipping tapers is probably the most ancient form of candle making still practiced on a regular basis today. Although time consuming, they are also amazingly easy to make. You need to use paraffin that has a melting point of 140 degrees, or pure beeswax to make these candles. An old coat hanger bent into a U shape with a hook at each end can make a good dipping frame. Have a place handy to hang the dipping frame from while the tapers harden. A hook or nail pounded partway into a wall works well for this purpose.

Choose a dipping vat that is fairly

deep. You can use a number 10, or gallon sized tin can, or an old stove top type coffee pot for the dipping vat. Place this dipping vat in a large saucepan, and put it on the stove top. Place chunks of wax in the dipping vat about as deep as you want the tapers to be long. Slowly pour water into the saucepan, making sure not to get any water in the dipping vat. Ideally the water should reach at least halfway up the outside of the dipping vat. Add as much water as you can to the sauce pan without causing the dipping vat to float or tip over. Turn the stove onto medium to start the water heating. You don't want it to boil wildly as that might cause the dipping vat to overturn. You want the water just hot enough to melt the wax and bring it up to the correct temperature. When the wax begins to melt add 5 Tbsp stearic acid per pound of wax, and put a candy thermometer in the liquid wax mixture. It may require over an hour for your wax to melt completely. You can speed the process up by melting the wax separately in a crock pot or wax melter, and pouring it into the dipping vat after it has reached the correct temperature. For smoothest dipping you want the wax temperature to be approximately 170 degrees F. As the wax continues to melt, prepare your wicks.

Cut a length of wick nearly 4 times the

desired length of your taper. Thread and wrap this wick around one hooked end of your dipping frame wire, across to, and around the second hook letting a length of wick hang down from each hook. Wrapping the wick around the hooks themselves helps prevent the wicks from slipping around on the wire dipping frame. Tie a small nut or washer around the bottom of each wick end to help hold the wicks straight.

The first time the wicks are dipped they should be held completely under the melted wax until the wicks are completely saturated, then allowed to cool. This primes your wicks to help prevent sputtering of your tapers as they burn.

Repeat dipping the wicks in one smooth, fluid motion, always dipping to the same point along the wicks. This technique may take a little practice to achieve the best product, but even a beginner can turn out a well-dipped candle with a little care.

Hang to let harden somewhat, then, dip again while still warm. It is important to maintain the temperature of your melted wax while this dipping process is being carried out. Repeat dipping and hardening until the tapers are the desired thickness. If you want a smooth finish on your tapers simply dip them in cool water after the final wax dip.

Trim the bottom of each taper with a razor while the candles are still warm, or use a fine toothed saw to trim after the tapers have completely hardened.

When completely hardened, unwrap the wicking from the wire frame and store the candles as-is, or prepare them for use by cutting the wick to desired length. Remember to store these candles in a cool place to avoid warping.

CHAPTER FIVE
BRAIDED RAG RUGS

THERE IS A CERTAIN BEAUTY in the look and feel of a home decorated with handcrafted items. Hand-braided rag rugs speak of thrift as well as pride in one's own skills and accomplishments. Using strips of material that might otherwise be destined for the landfill to create lovely throw rugs is probably one of the most economically and ecologically friendly endeavors a homesteader can engage in. I like to work on my rugs in the evening when all the other chores of the day are finished. I suppose that's because I find it so relaxing. It is also an easy project that can be completed in a relatively few evenings of work.

You can use whatever material you choose for rag rug making. You can save your own scraps, gather them from friends and family, or buy them if you can get fabric remnants at a good price. Sturdy woolens will make a more durable rug, but any fabric or fabric mix in a home-braided rag rug will last for many years. You can select the colors for your rug based on your color theme of the particular room you are making the rug for, or you can mix and match your colors for a fun, random patterned creation. You can achieve all sorts of colors and color variations by dying your material with fabric dye before making your rug.

To make your rug, begin by cutting your material into strips about two inches wide. Adjust the width of your strips based on the thickness of the fabric you are using. Thinner material should be cut wider, thicker material can be cut more narrow in order to achieve a uniform thickness for your finished braids.

Using three strips, knot them together at one end and beginning braiding them just as you would braid a child's hair. Keep the raw edges of the fabric folded in toward the center of the fabric strip to prevent your strips from fraying. Try not to get your strips tangled as you braid them. At first, you may find that working with shorter strips helps prevent the inevitable tangling. You want your braids snug, but not too tight. This will make it easier to join your braids later. Add new strips to lengthen your braid by sewing them onto the end overlapping the ends by about two inches.

When your braided strip is sufficiently long enough you can begin to form your rug. If you want a round rug simply roll the end of your braid into a tight loop and continue to wind it around in a flat spiral until the entire strip of braid has been used. To make an oval shaped rug just take a longer length of the braid and fold it back on itself continuing to wrap the braid around this long center piece in the same flat spiral form.

Sew the braid together using a darning needle and heavy thread. This is the easiest way to join the braid in the rug. Sew through the edge of one braid into the edge of an adjoining braid all around and throughout the rug until it is solidly joined.

There is also another way to join the braid of a rag rug. Save small strips of fabric that match the colors in your braids and sew them together to form a long thin strip. Use this to thread through the center of the braided rug material and into an adjoining braid, pulling firmly. This is where it pays to not braid your strips too tightly. If your braids do happen to be too snug to stick the small strip through with your fingers you can use a crochet hook.

When you come to the end simply use needle and thread to securely sew the end of the small strip into the larger braid, concealing it within the braid itself.

Rag rugs are machine washable, and can even be dried in the drier. It's best to tumble dry them only until still damp. Promptly remove the rug from the dryer and reshape if needed. Lay flat to finish drying. Shake your rag rugs outdoors to clean them between washings.

A wonderfully waterproof version of a rag rug can be created by following the above directions, but substituting plastic bread bags, which have been cut into strips, for the fabric strips. Braid and sew in the same manner, using nylon fishing line for the thread.

CHAPTER SIX
HOMEMADE YEASTS AND BREADS

HAVE YOU EVER WALKED into a bakery and smelled that luscious aroma of freshly baked breads and rolls still warm from the oven? Are you often tempted to buy bakery items because they taste so fresh and delicious even though they cost so much? Wouldn't you love to have that aroma and flavor come straight out of your own kitchen? Well, you can and for a fraction of the price it would cost you to get them from a bakery. You also will have the sense of pride in accomplishment you get from doing it yourself. Home-baked bread speaks to us of the warmth of hearth and home.

Flour, liquid, and yeast are the basic ingredients used in bread making. The flour you use should be all purpose, preferably unbleached, or higher gluten bread flour, or a combination of whole grains and one or both of these flours. The liquid used in bread making will usually be water, although some soft-textured breads call for either milk, or buttermilk, and a few even call for fruit juice.

Yeast is what leavens bread, or causes it to rise. Packaged active dry yeast is the kind most commonly used today, although old fashioned, homemade yeast is quite easy to make and imparts a rich, robust flavor to your breads. Even when you make your own yeast, you will still want to keep a supply of active dry yeast on hand to use with your home-made yeast if you want to achieve the light, tender breads we are so used to today. Many years ago when our great, great grandmothers baked their breads from sourdough or potato yeasts alone, they produced very flavorful, but very heavy and dense loaves. These old time yeasts, used together with the active dry yeast that is available now in any grocery store, will give you the texture we've come to appreciate in breads today combined with the more robust flavor and nutrition of yesterday.

Bread-making may seem like a dead or at least dying art to most homemakers today, but it actually is still very much alive. It does require a certain attention to details and some learned techniques, but these are things anyone can learn very easily.

Helpful Hints

1. Bread dough has to be kneaded for it to develop its characteristic tenderness. The more you knead your dough the lighter and airier your bread will be, because kneading activates the gluten in the flour. Kneading is accomplished by stretching, pulling, and folding your dough in on itself, then turning the dough a quarter turn and repeating the stretching, pulling, and folding motions. Repeat this process, working in more flour as needed until your bread dough is smooth and elastic rather than sticky. As a rule you'll have to knead the dough for 5 to 15 minutes to achieve this texture, depending on the type of flour used. All white flour dough needs a shorter kneading then does the whole grain flour dough.

2. Let your bread dough rise in an environment that is maintained at about 80 degrees F. The cooler the temperature the longer it will take for your bread dough to rise. If it is much warmer than 80 degrees F your dough will rise too quickly and the resulting bread will have large air holes in it. The warming compartment of a kitchen wood range is ideal for setting

bread to rise. The top of the electric or gas kitchen range works well too. Keep your bread pans of dough covered with a damp linen towel as the dough rises to prevent the top of the dough from drying out.

3. There are a few adjustments you need to remember to make when baking yeast breads at higher altitudes. Yeast doughs rise faster in higher altitudes and many breads have a coarser grain. To make a finer grained product use slightly less yeast then called for. Let your dough rise for a shorter time, just until barely doubled, and punch down the dough two times instead of once, letting dough rise one more time then called for in the recipe.

Yeast recipes

Sourdough starter

A tangy, robust yeast that is so good in so many breads. A treasure from yesterday that is still loved today.

1 package active dry yeast
1 ½ cups warm water (110 degrees F)
1 Tbsp sugar or honey
1 ½ cups all-purpose flour

Pour warm water into a medium-sized bowl. Sprinkle yeast over water. Stir in the sugar and let stand until mixture is foamy, about 10 minutes. Stir in the flour until the mixture is smooth. Cover the bowl and set aside at room temperature overnight. Then refrigerate starter at least 3 more days, stirring once each day. For each cup of starter removed, replace it with ½ cup flour, ½ cup warm water, and 1 tsp sugar or honey. This starter can be safely stored in your refrigerator for at least a month. It is the basis for many delicious breads, biscuits, and pancakes.

Potato Yeast

A wonderful old fashioned yeast that is very easy to make.

2 quarts water
4 large potatoes
1 cup sugar
2 tbsp salt
1 package active dry yeast

Peel the potatoes and soak them in 1 quart of water for about half an hour. Put the other quart of water into a medium saucepan over medium-high heat, and bring it to a boil. Grate the potatoes quickly so they don't discolor, and put them into the boiling water, stirring and cooking for 5 minutes. Remove from heat, and add the sugar and salt. Stir well, and turn into a glass or plastic bowl or jar, and let stand until lukewarm. Add the yeast and mix well. Cover the bowl or jar loosely, and leave out to ferment at room temperature for about 3 ½ hours. Cheesecloth or a linen towel works well as a covering for this stage. Stir the yeast down every time it rises to the top of the container. Transfer the yeast to a jar with a tight fitting lid and put it in your refrigerator. It will keep for up to two weeks in the refrigerator. The next time you want to make this yeast simply save a cupful of this starter and substitute it for the active dry yeast in the recipe above.

Bread Recipes

Sourdough Bread

The famous bread with the tangy flavor, dense texture, and hearty crust that is sure to please everyone in your household.

1 package active dry yeast
1 ¾ cups water
1 cup sourdough starter
1 Tbsp sugar
7 cups all purpose flour
1 ½ tsp salt
1 ¼ tsp baking soda
1Tbsp vegetable oil

In a large bowl, sprinkle yeast over the warm water. Stir in the sugar and let the mixture stand until foamy, about 10 minutes. Then add the sourdough starter, 5 cups of the flour, the salt, and the baking soda to the yeast mixture. Beat with a wooden spoon until mixture is smooth. Beat in enough remaining flour to make a soft dough. Turn the dough onto a bread board or a lightly floured surface. Knead the dough, working in more flour as needed, until the dough is smooth and elastic, 5 to 10 minutes. Form dough into a ball. Put the vegetable oil in a large bowl. Transfer the dough to this bowl and turn the dough once to bring the oiled side up. Cover with a clean, damp cloth. Let the dough rise in a warm place, free from drafts, until it has doubled in bulk, about 45 minutes. Grease a large baking sheet. Punch down the dough. Shape it into one large or two medium sized round loaves. Place the dough on the greased baking sheet. With a sharp knife make 3 slashes across the top of each loaf. Cover the dough with a clean, damp cloth and let rise in a warm place until double in bulk, about 45 minutes. Heat the oven to 400 degrees F. Bake bread for 25 to 35 minutes, depending on loaf size, or until each loaf sounds hollow when tapped on the top. Remove from baking sheet and cool on a wire rack.

Red Raspberry Sourdough Griddle Cakes

A delightful change for a Sunday morning brunch!

1 ½ cups all purpose flour
3 Tbsp sugar
1 tsp salt
½ tsp ground allspice
¼ tsp baking soda
1 cup milk
1 large egg
2 Tbsp vegetable oil
1 cup sourdough starter
1 ½ cups red raspberries

In a large bowl, mix the flour, sugar, salt, allspice, and baking soda. In a small bowl, combine the milk, egg, and oil. Stir the milk mixture and the sourdough starter into the flour mixture just until well blended. Gently fold in the raspberries. Lightly oil your griddle; heat it over medium-high heat until a drop of water will dance across the surface of the heated griddle. Spoon 1/3 cup batter onto the griddle for each cake. Cook until the top is covered with bubbles and the edges are dry. Turn and cook the other side for 1 or 2 minutes. Add more milk to the batter if it is too thick to spread well when put on the griddle, and adjust the heat as needed. These griddle cakes are delicious served with butter and warm maple syrup.

Delicious Whole Wheat Bread

8 cups whole wheat flour
1 package active dry yeast
3 tsp salt
1 cup potato yeast
1 ½ cup very warm water
2 Tbsp vegetable oil
1 egg white
1 tsp water
1 tsp sesame seeds
1 tsp sunflower seeds

In a large bowl combine 6 cups of flour, active dry yeast, and salt. Add the potato yeast, warm water, and oil. Beat this mixture together until a soft dough forms. With a wooden spoon, stir in enough remaining flour to make a stiff dough. Turn the dough onto a lightly floured surface. Knead dough, working in more flour as needed, until the dough is smooth and elastic, about 10 minutes. Shape dough into a ball and let it rest 5 minutes. Lightly oil a large bowl, and transfer the dough to the bowl, turning once to bring oiled side up. Cover bowl with a clean, damp cloth and place it in a warm place to rise until doubled in bulk, about 45 minutes.

Grease 2 loaf pans. Punch down dough and shape into 2 ovals. Place in greased loaf pans and cover with a clean, damp cloth. Let rise in a warm place until doubled in bulk, about 20 minutes.

Preheat oven to 400 degrees F. In a measuring cup, combine the egg white with the water, and brush this over the top of the loaves. Sprinkle with sesame seeds and sunflower seeds. Bake loaves 30 to 35 minutes, or until they sound hollow when tapped on top. Remove from pans and cool on wire rack. For a softer crust, drizzle melted butter over warm loaves. Makes 2 loaves.

Basic White Bread

1 package active dry yeast
1 ½ cups warm water
¼ cup sugar
5 ½ cups all purpose flour
1 tsp salt
3 Tbsp vegetable oil

In a large bowl measure 4 cups of the flour. Sprinkle yeast on top of the flour and add the sugar and salt. Mix in the water and vegetable oil. Beat with a wooden spoon until mixture is smooth. Stir in enough remaining flour to make a soft dough. Turn dough out onto lightly floured surface and knead well for 5 to 10 minutes, adding more flour as needed until dough is smooth and elastic. Lightly oil a large bowl and transfer dough to the bowl turning once to bring oiled side up. Cover bowl with a clean, damp cloth and set aside in a warm place to rise until doubled in bulk, about 45 minutes.

Grease a 9x5x3 inch loaf pan. Punch down dough and shape it into an oval. Place dough in the greased loaf pan and cover the pan with a clean, damp cloth. Let dough rise again in a warm place until doubled in bulk, or about 45 minutes. Heat oven to 350 degrees F. Bake the loaf for 50 to 60 minutes or until it sounds hollow when tapped on top. Remove the loaf from the oven and let it cool on a wire rack. For a softer crust, brush the top of the warm loaf with melted butter.

Old fashioned oatmeal bread

A delicious, heart-healthy bread

2 cups milk
2 cups quick rolled oats, uncooked
¼ cup brown sugar, firmly packed
1 Tbsp salt

2 Tbsp vegetable oil
1 package active dry yeast
½ cup warm water
5 cups all-purpose flour
1 egg white
1 Tbsp water
Rolled oats

Scald the milk in a large saucepan over medium heat. Stir in 2 cups of rolled oats, brown sugar, salt and vegetable oil. Remove from heat and allow to cool to lukewarm.

Sprinkle yeast on warm water, stirring to dissolve.

Add yeast and 2 cups of flour to milk mixture. Mix vigorously with a wooden spoon for 3 minutes or until batter is smooth. Add enough remaining flour, a little at a time, to make a soft dough that cleanly leaves the sides of the bowl. Turn out onto lightly floured surface. Knead until dough is smooth and elastic, 8 to 10 minutes. Lightly oil a large bowl and place dough in bowl, turning once to bring oiled side up. Cover with a clean, damp cloth and put in a warm place to let rise until doubled in bulk, 1 to 1 ½ hours. Punch dough down, and let rise again until nearly doubled in bulk, about 30 minutes. Turn dough onto lightly floured surface and divide in half. Round up to make 2 balls. Cover dough with inverted bowl and let dough rest for 10 minutes.

Grease two 9x5x3 inch loaf pans. Shape dough into loaves and place in the loaf pans. Cover pans with clean, damp cloth and put aside in a warm place to rise until almost doubled in bulk, about 1 hour. Beat the egg white with the 1 Tbsp water. Brush tops of loaves with this egg white mixture. Sprinkle tops with rolled oats. Bake loaves in 375 degree F oven for about 40 minutes or until done. Loaves will sound hollow when tapped on top when done. Remove from pans and cool on wire racks. For a softer crust, brush with melted butter while still warm.

Note: You can modify this recipe by replacing ½ cup of the milk with ½ cup of potato yeast in the above recipe. This will make a lighter dough and a more tender bread.

CHAPTER SEVEN
HOMESTEAD DAIRY

Fresh Butter

FRESH BUTTER IS SOMETHING all of us can enjoy whether we own a cow or not. Perhaps you are lucky enough to live near a dairy or co-op where you can obtain heavy cream at a reasonable cost. You very well may have to go to the grocery store to buy your cream, but one taste of your own homemade butter and you'll agree that it is well worth the effort. If you can afford to invest in a hand crank butter churn you will never be sorry. The jar itself is usually made of glass with a hand crank churn assembly seated on the screw-top lid. If you cannot find, or cannot afford one of these old time butter churns, you can do just as good a job using a quart size mason jar, a clean marble, and a little elbow grease. Pour about a pint of very cold heavy cream into the jar and drop in a clean marble. Screw on the lid and band, or another screw top lid that fits, and simply shake the jar.

It will take a great deal of shaking to make butter this way, between fifteen and thirty minutes, but children usually love to help with this homestead chore. When the butter is nearly ready, it will be seen as yellow flecks floating in the milk. Soon the butter will clump solidly together. When this happens it is time to rinse the butter clean. Drain the buttermilk into a storage container for later use, and refrigerate it. Then collect all the butter into one lump and put it in a bowl set under cold, running water. Simply knead the butter under the cold water working out as much buttermilk as possible. This is an important step that should not be overlooked because any buttermilk left in your butter can lead to faster spoilage and result in an off flavor in your finished product. When the water runs clear and all the buttermilk has been rinsed away, it is time to salt your butter. You don't need to use very much salt to make your butter pleasing. Taste the butter as you add small amounts of salt to achieve the right amount of salt for your own tastes. Too little salt is better than too much. You really don't have to add any salt at all if you prefer unsalted butter, however unsalted butter does not retain the fresh flavor as long as salted butter does. Store your butter in a tightly covered plastic container in the refrigerator.

Home Cheese Making

You can make delicious cheeses quite easily right in your own home. Most soft cheeses can be made using equipment already found in most homestead kitchens. Harder cheeses require a bit more advanced equipment but will be worth the initial outlay in funds. The savings over buying commercially processed cheese will repay you for all of the financial investment you have to make.

The basic ingredients in cheese making are milk, starter culture or natural acids, and rennet.

Milk makes the cheese

The most important ingredient in cheese is the milk. Cheese can be made from just about any kind of milk including, but not limited to, cow, goat, sheep, and mare. Each type of milk has its own unique characteristics that make it ideal for certain types of cheese. Are you one of the lucky homesteaders who have access to your own fresh cow's or goat's milk? Perhaps you have access to a local dairy or farmer's co-op where you can purchase your milk. Yet, even if you have to purchase your milk from the local supermarket, you can still make quality cheese at a fair price.

Milk is mostly water. Cow's milk contains 87 percent water by weight. Remove that water and what remains are the basic components of cheese; fat, protein, lactose, and minerals. Cheese making begins with removing the water from the milk. This can be accomplished in a couple of ways, both of which rely on the addition of an acid to the milk. The acid causes the milk protein to coagulate into curd which is the solid protein of milk. The simplest method is to add an acid like lemon juice or vinegar to the milk. Another more preferred method is to add bacteria, which creates an acid in the milk. The addition of bacteria also enhances flavor and adds character to the finished cheese.

Store bought milk has been homogenized, which is a process that breaks up the fat globules to such a tiny size that they are suspended in the milk and will never separate the way fresh milk does. Homogenizing milk also alters the milk protein, and unless you do something to compensate for this the milk will never make a satisfactory curd for hard cheese. Even soft cheese made strictly from this milk alone will have a wax-like texture that sticks to the teeth and hinders flavor development. The easiest way to fix this problem is to use the lowest fat milk available and replace the needed fat with heavy cream or whipping cream. Although the cream has also been homogenized, the fat ratio is much higher and produces a much better product.

You can even use powdered non-fat milk combined with heavy cream or whipping cream to make a very good cheese. Reconstitute the dry milk using the directions on the box, and add 1 pint of cream per gallon of milk and use this in place of whole milk in most cheese recipes.

An alternative to this is to use calcium chloride to compensate for the processing of store bought milk. The addition of calcium chloride will help restore the milk protein and aid in the development of a quality curd.

Cheese starter cultures

The acidification of milk in cheese making using bacterial cultures is called ripening. The bacterial culture itself is called a starter. Starters are added to the milk after the pasteurization process and at specific temperatures. They are allowed certain time periods to work undisturbed, depending on the type of cheese being made. During this time period the bacteria consumes the lactose, or milk sugar. As the lactose is consumed lactic acid is produced which causes the milk protein to develop into the curd. Other byproducts of this ripening stage produce the individual flavor and characteristics of the finished cheese.

There are two basic forms of cultured starters for cheese making. Thermophilic, which is a heat loving starter, and mesophilic, which is a moderate temperature starter. Cultured buttermilk is an example of a mesophilic starter. Yogurt is an example of a thermophilic starter. There are also many variations on these starters. As you improve in your cheese making skills you will want to learn to culture your own starters. However, as a novice cheese maker it will be easiest for you to purchase your *mesophilic* starter which can be purchased at any cheese making supply store, or can be ordered online. This comes in a powdered form in a small packet similar to bread yeast. It is best to purchase what is known as direct vat inoculation, or DVI. This is added directly to the milk with no pre-culturing necessary.

Kid lipase powder

Lipase is an enzyme used for the development of the distinctive flavors of certain cheeses. This enzyme is a vital ingredient in the manufacture of cheeses such as Feta, Romano, Parmesan and mozzarella. Without adding lipase your cheese will never achieve the distinctive flavor that each of these cheeses are known for.

Cheese making rennet

Rennet is an enzyme which acts on milk protein and causes it to coagulate. Rennet is available in a liquid as well as tablet form. It does not have a long, stable shelf life, so do not attempt to stock up on rennet. It is best to have fresh rennet to ensure the best results. One teaspoon liquid rennet equals one rennet tablet in most cheese recipes.

Salt

Use only non-iodized salt, such as kosher or canning salt, in cheese making. Iodized salt will give your cheese an unappealing, greenish tinge.

Cheese making equipment

All equipment used in cheese making must be made of glass, plastic, enamel lined steel, or stainless steel. Due to the acidification process used in cheese making, the use of aluminum and cast iron utensils is not appropriate. The acidity of the milk can cause these metals to leach into your cheese.

You will need:

- A stainless steel or enamel lined pot that holds at least ten quarts, and a secure fitting lid for this pot.
- A long-bladed stainless steel carving knife that is long enough to reach to the bottom of your cheese pot.
- A floating dairy or candy thermometer.
- Stainless steel or glass measuring cups.
- A couple of long, stainless steel cooking spoons.
- Stainless steel measuring spoons.
- About 2 yards of heavy cheesecloth or other open-weave material.
- A large stainless steel or plastic strainer.
- Cheese drainer/form for hard cheese only.
- A cheese press for hard cheese only.*
- Cheese wax for hard cheese only.
- Waxing brush for hard cheese only.

*If you don't want to spend the money on a new cheese press, you can make your own with an empty coffee can. Cut out both ends of the can and then cut out a circular wooden piece just large enough to fit in the can opening. Take a couple of small bricks, wrap them in foil and use these to press the wooden circle into the can after the cheese is inside it.

You must be sure all utensils are clean and sanitized before each use. If your dishwasher has a sanitize cycle you can use that. You can use a weak bleach and water solution to sanitize your plastic and glass utensils, as well as your countertops, however, bleach should never be used on stainless steel as it will cause the stainless steel surface to deteriorate. Boiling your utensils for 10 minutes is also an effective way to sanitize them. There are also various chemical sanitizers available through cheese making suppliers if you choose to use them.

Aging your cheese

After your hard cheeses are waxed they can be aged in several different ways. The ideal temperature for aging cheese is around fifty five degrees F.

A basement or root cellar works great for this purpose, however you must be vigilant in ridding the area of all rodents. Once mice get into your cheese, you will have to throw it away. A large picnic cooler is ideal for keeping your aging cheese safe.

Un-waxed cheeses can be kept for several months if covered with oil. Any vegetable oil will do nicely, but you need to be sure that the entire cheese is completely immersed in the oil. Mold needs air to grow, and it will find any tiny area that was overlooked in the dipping process. It is much better to have a little too much wax than to not have enough and have your cheese mold as a result.

Cheese Recipes

Homemade Cream Cheese

This rich cream cheese is surprisingly simple to make and delicious to eat!

6 cups milk
4 cups heavy cream
1 cup plain yogurt
1 rennet tablet
¼ cup cold water

In a large, non-aluminum saucepot, combine the milk, cream, and yogurt. Heat the mixture over low heat, stirring occasionally, until the mixture reaches 100 degrees F, about 30 to 45 minutes.

In a glass measuring cup, dissolve the rennet tablet in the water. Add this to the warmed milk mixture. Stir just until the mixture becomes thick and creamy. Remove the saucepot from the heat. Cover the pot and let it stand in a warm spot in the kitchen until the curd separates from the whey, 1 to 1 ½ hours. If the temperature of the curds drops below 85 degrees F during this time, simply reheat the cheese slowly.

Cut the curd into 1-inch cubes, and let these stand undisturbed for 15 minutes. Line a large glass or stainless steel strainer with several layers of damp cheesecloth. (Dampening the cheesecloth helps keep it from slipping around on the strainer.) Place this over a large bowl to catch the whey as it drains off the curds. Pour the curds into the lined strainer and put the bowl and strainer assembly in the refrigerator. Allow the curds to drain overnight.

The next day, discard the whey and remove the curds from the cloth. This cheese is now ready and can be used immediately as it is, fresh or in baking, or you can add ½ to 1 tsp salt, to suit your tastes, and wrap it in plastic wrap. Refrigerate up to 1 week.

Quick and easy cottage cheese

The perfect cheese for the beginner cheese maker.

1 gallon whole milk

½ cup cultured buttermilk
¼ tsp liquid rennet or ½ rennet tablet
¼ cup cool water

Warm the milk to 86 degrees F. Stir in the buttermilk. Dissolve and mix rennet with water, and add it to the warmed milk. Set aside until it coagulates, usually about an hour. Cut the curds into ½ inch cubes. Heat slowly over low heat until the temperature reaches 110 degrees F. Hold curds at this temperature for 30 minutes, stirring often to prevent matting.

Line a plastic or stainless steel strainer with several layers of damp cheesecloth. Put curds into this cloth lined strainer and let drain for 20 minutes. Lift the curds in the cheesecloth and dip them in a pot of cold water, then let drain until the curds stop dripping. Place the cheese in a bowl and add salt and cream to taste. Store unused cheese in the refrigerator, in a tightly sealed container, for up to 1 week.

Cheddar Cheese

A little harder to make and a little longer to wait for the final product, but this delightfully aged cheese is well worth the effort and the wait!

1 ¾ gallons whole milk

1 quart heavy cream
½ cup cultured buttermilk or substitute
¼ tsp mesophilic DVI
1 tsp liquid rennet or 1 rennet tablet
½ cup cool water
4 tsp salt

In a large stainless steel or enamel pot, mix milk and cream together. Warm milk mixture to 88 degrees F. Stir in buttermilk or mesophilic DVI. Set the milk aside to ripen in a warm place for 1 hour. Do not let the temperature of the milk fall below 88 degrees F during the ripening process. Dissolve and mix rennet in cool water and stir into the milk. Bring the temperature of the milk mixture back up to 88 degrees and maintain at that temperature for another 45 minutes to coagulate the milk. The curd is ready to cut when you stick your finger into the curds and they break cleanly around your finger and the hollows fill with whey. This is the clean break stage. Cut the curds into ½ inch cubes and let them rest for 20 minutes. Slowly bring the temperature of the curds up to 98 degrees F, stirring gently. Maintain the 98 degree F temperature for 30 to 45 minutes or until the curds are firm and somewhat dry, stirring often to prevent the curds from matting together. Remove from heat, and let the curds settle to the bottom of the pot. Line a plastic or stainless steel strainer with several layers of damp cheesecloth. Carefully pour off some of the whey, then pour the remaining curds and whey into the cloth lined strainer, and let drain for 10 minutes. Place the curds back into the pot and stir in the 4 tsp salt. Mix this up well, breaking up any curds that have matted together. Keep the curds in the pot in a sink full of warm water for one hour. Stir often to prevent matting. Line your cheese press with cheesecloth leaving some cloth hanging outside the press. Spoon the curds into the cheese press and fold the extra cheesecloth over the top of the cheese. Place the wooden follower on top of the cheese and press at 15 pounds of pressure for 20 minutes. Remove the cheese from the press. Again, put clean cheesecloth in the press leaving extra hanging out the end. Unwrap the cheese and turn it over, putting it in the press opposite the way it was the first time. Fold the cheese cloth over the cheese and press at 30 pounds pressure for two hours. Remove the cheese from the press again. Redress your press the same way, unwrap your cheese and put it in the press again, folding the ends over the cheese. Press at 35 pounds pressure overnight. In the morning, remove the cheese from the press and allow it to air dry for several days until it is dry to the touch. Turn the cheese several times each day while it is drying. Coat with cheese wax after the cheese is dry to the touch. Be sure to completely coat the cheese with the wax. Age at 55 degrees F for at least two months. The longer you age the cheese, the sharper the flavor your cheese will develop. The best cheddar is aged for 12 months or longer.

Feta type cheese

Feta is a semi-hard, aged cheese that is made from the milk of sheep, cows or goats.

2 gallons milk
¼ tsp mesophilic DVI culture
¼ tsp kid/lamb lipase powder
½ tsp liquid double strength rennet
½ cup cool water
Kosher salt
Brine

In a stainless steel or enamel double boiler warm milk to 86 degrees F. Add the starter culture and the lipase. Stir well, and let ripen in a warm spot for one hour. Keep the milk at 86 degrees F. throughout this time period. You can put your covered pot of heated milk in a cloth lined, Styrofoam cooler to help maintain the temperature of the milk mixture.

Dissolve the rennet in the water, and add to the milk mixture, stirring briskly for 5 minutes. Cover the pot and let set about 40 minutes. After this time, check the curd by poking your finger into the curd. If the curd breaks cleanly around your finger and whey fills in the hollows, it is at the clean break stage.

Cut the curd into ½ inch cubes. Let the curd rest for 10 minutes. After 10 minutes, stir the curd gently, cutting any curd that did not get cut the first time. Cover, and set the curd aside for 30 minutes, keeping the curd at 86 degrees F, stirring occasionally to keep the curd from matting.

Line a stainless steel or plastic strainer with several layers of damp cheesecloth. Place the cloth lined strainer over a large bowl in the sink. Carefully pour the curd into the strainer. Lift the four corners of the cheesecloth and tie them together. Hang this bag to continue draining. You can hang it over your kitchen faucet for convenience.

Put the whey in a container for use in the brine, or for making ricotta cheese later. Let the cheese hang for 24 to 48 hours. It will develop a particular, cheesy odor. After this time, remove the cheese from the bag, and cut the cheese into 2 or 3 inch cubes. Sprinkle the cubes on all sides with kosher salt. Place the cubes in a large, tightly covered glass or plastic container. Let the container of cheese sit at room temperature for 2 to 3 days to harden up the cubes.

Make the brine by mixing ¼ cup coarse, kosher salt with ½ gallon of water or whey. Add the brine to the cheese cubes. Keep the cheese in the refrigerator. This Feta cheese will stay preserved in this brine, when refrigerated, and will continue to age for up to a year.

If your Feta develops any moldy spots, simply remove the moldy part and discard it. The rest of the cheese is still perfectly fine to eat. If your Feta gets soft in the brine it only means the cheese wasn't hung quite long enough. The cheese is still good, however because of the softer texture you will want to use it in cooking rather than for fresh use. Simply pour off the brine and use the cheese within a few weeks.

Ricotta Cheese

Don't throw away that whey! Turn it into ricotta cheese and make some lasagna or cheese stuffed shells, or ravioli. It's so easy to make. Simply save all whey from one day of cheese making. Strain it through a couple layers of cheesecloth to remove all larger bits of curd, or your ricotta will develop rubbery globs in it. After straining, pour the whey into a non-reactive pot or kettle of a sufficient size. Heat to nearly boiling, or about 205 degrees F, stirring often to prevent scorching.

Remove from heat, and allow to cool on its own to lukewarm. Pour whey off into a strainer that has been lined with a piece of close-weave material like a non terry dish towel or a clean white pillow case. Allow all whey to drain off. This will take several hours. Ricotta is a very small curd cheese by product, and the tiny curd will drain slowly. When sufficiently drained, pack ricotta into airtight containers and refrigerate for up to a week, or freeze for future use. You can catch the whey that drains off the ricotta and use it in bread making or in soups. It is very high in riboflavin and adds a delightful, full bodied flavor to hearty meat soups and stews.

Yogurt

Yogurt is a fermented milk product, similar to cheese. It is easily digestible and is known to replenish non pathogenic flora in the gastrointestinal tract. Yogurt is not difficult to make, especially with the electric yogurt makers that are available today. Yogurt can easily be made with only the items found in the average homestead kitchen.

The same basic principles that apply to cheese making, also apply to making yogurt.

Sanitize your utensils by boiling them in water, or with special chemical sanitizers that are available for this use. A weak water and bleach solution can be used to sanitize glass or plastic utensils and countertops. Bleach should never be used on stainless steel as it will cause the surface of the metal to deteriorate.

Equipment

- A 2 ½ quart double-boiler with lid
- Yogurt maker

If not using a yogurt maker:

- 9 eight ounce mason jars with lids
- Cheese or candy thermometer
- Styrofoam cooler

Fruit on the bottom yogurt

You can use low-fat, skim or whole milk in this recipe, however, whole milk makes a richer finished product.

½ gallon milk
½ cup plain yogurt, live culture for starter
14 Tbsp prepared fruit or fruit pie filling

Heat the milk to 190 degree F. Keep covered. Remove from heat. Place covered pot in a pan of clean, cool water, stirring occasionally until the milk is about 130 degrees F.

Put 2 Tbsp of desired fruit or fruit pie filling into the bottom of each jar, leaving one jar empty. Stir up the yogurt starter with a clean fork. Add this to the milk when it reaches the 130 degree F mark, stirring thoroughly. This should drop the temperature to about 120 degrees F. Immediately pour this mixture into the jars that contain the fruit, and finally, fill the empty jar. This will be the new starter jar. Cover the jars immediately with sterile lids. Place the jars in yogurt maker and follow manufacturer's instructions. If not using a yogurt maker, place jars in the Styrofoam cooler. Add enough warm water, 122 degrees F, to the cooler so that the jars are surrounded, but the water is well below the lid rims. Do not disturb the yogurt, but keep an eye on the water temperature, adding more 122 degree F water if the temperature drops below 105 degrees F. Your yogurt will be finished in about 3 hours. Refrigerate your yogurt until needed.

CHAPTER EIGHT
HOME FREEZING

THE GARDEN IS OVERFLOWING with vegetables and the orchard trees hang heavy with fruit. The birds are already helping themselves to your cherries and blackberries. You also have a fat calf ready for slaughter, and your chickens seem to be working overtime. You know that the cold winds of winter will be blowing in just a few short months. So how do you preserve this abundance of food for your family's use? Freezing is one of the easiest methods of preserving all that bounty.

A freezer is no longer considered a luxury item. It is a vital appliance for proper food storage in today's homestead.

Freezing preserves food by delaying or stopping the growth of bacteria, molds and yeasts and delaying the actions of enzymes. Freezing does not kill these microorganisms or enzymes the way canning does.

Some foods items, strawberries for example, simply do not hold up well under the processing required to can or bottle them for home storage. However, these same foods can be frozen very successfully, and their thawed flavor will be very close to that of the fresh item. Peaches are another food that freeze as well or better then they can. The difference between frozen peas and canned peas with regards to flavor, texture, and color is also very noticeable. Quite simply, the rewards of owning a freezer when you have access to quality homegrown produce, dairy products, and meats is unlimited.

Freezers can be purchased for very little money these days, especially if you can pick one up used. Be sure that you see it run before you buy it. Feel to see that it gets cold enough to actually freeze your foods. Check the door gasket to be sure that it seals properly and there are no costly air leaks.

Chest type freezers generally hold more per cubic foot then do upright freezers, however, with an upright freezer you probably won't be digging through a year's supply of food to get to that one certain item on the bottom. Upright freezers also come in a frost free version that chest types don't offer. The biggest drawback to the upright freezers will be their cost. They simply cost more than the chest type, but many people find the benefits of the upright outweigh the added cost.

You can calculate how big a freezer you will need based on how much food you go through on an average year. For example; a 9 cubic foot freezer will hold about 315 pounds of food. A 15 cubic foot freezer will hold approximately 525 pounds of food. A 22 cubic foot freezer will hold 770 pounds of food. Remember though, that your freezer will hold more than this in any given year due to food turnover rate.

Freezer storage times vary depending on food type. As a general rule, fruits and vegetables have a maximum storage time of 8–12 months. Beef and poultry should not be stored in the freezer for more than 12 months. Pork should be used within 3 months of freezing, and lamb and veal can be stored up to 9 months. Big game can be stored in the freezer for up to 9 months. As with any food storage system rotation is vital, therefore everything that goes into the freezer must be identified and dated on the label.

Before you purchase a freezer consider the space you will put it in and buy one that is not too big for the space allowed. Freezers need a little space to breath, and will function most economically when they are not cramped into a tiny space. Do not use your chest type freezer for an extra work surface or storage area. Any extra weight on the freezer lid will distort the lid gasket making it more susceptible to air leaks, which can be costly both in energy wasted as well as in food quality deterioration.

Care of your freezer is a simple task. Keep the outside clean and dry, and defrost the inside of non-self defrosting models once a year. Even self defrosting models should be washed and inspected once a year. This is also a good opportunity to take an inventory of foods in the freezer, moving foods that need to be used sooner to the top or front of the freezer unit. If you can put your frozen food into a couple of coolers your task is much easier. You can put all your frozen food into cardboard boxes that are then wrapped in blankets and put in a cool part of your home, Work quickly on defrosting your freezer. Turn off electricity and unplug the freezer. Remove all ice buildup on the sides and bottom of your freezer. You can scrape it off with a plastic or wooden scraper, or use warm water to melt it off in sheets. Never use metal to scrape your freezer's interior or exterior surfaces. You can also put a large pan of hot water inside the freezer, and close the lid. Clean up any spilled food on the interior of your freezer, using a scraper if necessary. Wipe the inside with a cloth that has been wet with a solution of 4 Tbsp baking soda in 1 quart of warm water, then dry all surfaces completely. Plug the freezer back in and turn on the electricity. Let the freezer cool for about 10 minutes, and return the food to the freezer.

As in any other food preparation principles, cleanliness must be observed to avoid contamination of foods.

Success in freezing depends largely upon proper packaging. In the extreme dry cold of the freezer, even solidly frozen foods can lose moisture. All foods must be protected from this environment by proper wrapping materials, and moisture and airproof containers. Your wrappings must also be tight to the food, as airtight as possible, because air draws moisture from the food resulting in freezer burn. Many foods are sensitive to oxygen, such as meats, poultry, and fish, and may become rancid from exposure to the air.

Proper wrapping materials include waxed freezer paper, plastic freezer bags, and vacuum sealed plastic freezer bags, waxed cardboard freezer boxes, wide mouth glass canning jars, and re-usable plastic containers.

Do not use waxed paper, storage type plastic wrap, regular aluminum foil, paper bags, or storage type plastic bags. None of these will adequately protect the foods in the freezer from freezer burn and deterioration.

Label your packages clearly using a wax pencil or heavy black crayon. Always include information such as the date the food was frozen, name of product, and number of servings in the container.

Basics on Fruits and Vegetables

Choose fruits and vegetables that are at the peak of perfection for freezing. Fruits and vegetables that are ripe for the table are ripe for the freezer. Do not let your fresh fruits and vegetables wait in the kitchen any longer then is absolutely necessary. Freezing them as soon as possible results in better textures and retained flavors. You also should pay attention to the specific varieties that are recommended for freezing. Remember, a freezer will not improve the quality of food you put into it, but it will preserve the quality and prevent spoilage of the food you put into it. The quality you put in is exactly what you will get out.

Prepare your produce quickly once it has been brought in from fields or orchard. Sort, wash, and prepare it according to recipe directions. Do not let it set in the store room or kitchen for any length of time. Remove over-ripe produce and discard. Set aside under-ripe produce for further ripening.

Wash all produce in cold water to remove dirt and any pesticide residue that may be on it.

Fruits

All fruits are not created equal. Some fruits can and bottle well, others do not. Most fruits freeze well, with very few exceptions. Pears are one of these exceptions. The quality of pears, both in flavor and texture, deteriorates in the freezer. They are simply better bottled or canned then frozen.

Over-ripe bananas can be frozen right in the peels for future use in banana bread and banana muffins. Just wash a couple of whole, ripe bananas and put them in a freezer bag, label and freeze.

Strawberries are a delicate fruit, and they do not can or bottle well. They freeze wonderfully, however, and this is the preferred method of preserving them. Simply wash the berries, and remove the caps. You can slice them or leave them whole. You can add sugar to taste and freeze them sweetened, or freeze them unsweetened and sweeten them after they are thawed, depending upon their use. Strawberries freeze well in re-sealable plastic freezer bags. After the berries are sorted and washed, slice and place them in the desired size bags and freeze them. It's that easy.

Blueberries, blackberries and raspberries also freeze the same way, either sweetened or unsweetened. If you sort and gently wash these berries, then spread them on baking sheets to dry, you can then place them in the freezer bags with no moisture on them. When they freeze they will freeze free, like marbles, and you can easily separate them to take only a cupful out of the bag at a time for pancakes and the like. Of course you do not slice blueberries, blackberries or raspberries when freezing them. They are just put into the bags whole.

Most fruits freeze and thaw very well without added sugar and are more versatile when frozen unsweetened. If you want to freeze your fruit sweetened there are several methods of doing so. One method is to freeze them in dry sugar. Spread the fruit in a single layer in a shallow baking pan or cookie sheet. Sprinkle light fruit such as peaches and apples with a mixture of 2 Tbsp lemon juice in 1 quart of cold water. A spray bottle works well for this as it will uniformly wet the fruit.

Sprinkle with ½ cup sugar per quart of fruit for most fruits. Increase sugar depending upon acidity of fruit being frozen. Gently turn fruit over and over until the sugar dissolves and juice is formed. Fill your freezer bags packing fruit as tightly as possible without crushing. Add more fruit as needed. Seal bags, label and freeze.

Another method of sweetening fruit for the freezer is to freeze it in syrup. Mix 3 cups of sugar to 1 quart of water in a saucepan, and bring to a boil, stirring constantly until sugar dissolves. Refrigerate until syrup is cold. Pour about ½ cup of ice-cold syrup into each canning jar or pint size freezer container. Fill jar half full with desired prepared fruit. Shake the jar gently to pack the fruit down as much as possible without crushing the fruit. Add as much fruit as possible and fill the rest of the jar with additional syrup to cover the fruit, leaving ¾ inches head space per pint. A small piece of crumpled foil or wax paper can be gently pushed down into the top of the fruit to hold it under the syrup if desired. Wipe top of jars and threads with a clean, damp cloth, and screw caps on. Label jars and freeze immediately.

Peaches are equally good; bottled, canned, or frozen. I think frozen peaches retain a bit more of the fresh peach flavor, so that is my preferred method of preserving them. Try it and you'll see. There is nothing so good as freshly thawed peaches smothered in whipped cream over shortcake. No canned peaches can come close to the nearly fresh flavor of frozen peaches.

To freeze peaches, sort, wash and peel the fruit. To peel peaches the easy way, put 5 or 6 peaches in a wire basket or cheesecloth and dip them in boiling water for 15 to 20 seconds, or until the skins loosen. Then plunge them into ice cold water and pull the skins off, and cut them in half, removing the pits and discarding. You can leave the peaches in halves, or you can cut them in slices. To prevent the peaches from turning brown after they are thawed, you must treat them before they are frozen with a solution of ascorbic acid. To do this, put the peach halves or slices into a bowl containing 1 gallon of water into which has been dissolved 1 Tbsp of ascorbic acid powder. Ascorbic acid can be obtained under several different name brands. Look for color preserving agents like *Fruit Fresh* in the canning and freezing section of your supermarket. You can crush vitamin C tablets if you cannot find ascorbic acid powder. You can also add ¼ cup of lemon juice or vinegar to 1 gallon of water to achieve the same result. Let the peaches soak in this solution until you have finished peeling, pitting, and slicing that entire batch of peaches. Pick handfuls of peach halves or slices out of the solution, letting the excess water drain out through your fingers back into the bowl. Gently pack the peaches in freezer bags, label, and freeze. You can freeze peaches sweetened if desired.

Freezer Jams

Open a jar of homemade strawberry freezer jam on a cold January morning, and it will feel like you have opened a jar of warm, June sunshine. No bottled or canned jam, especially the store-bought kind, can come close to the fresh-picked flavor of your own freezer jam. Strawberries, raspberries, blackberries, blueberries, and peaches all lend themselves wonderfully in making delicious freezer jam. The ingredients are few, and the preparation is simple. Fresh fruit, pectin, a little lemon juice, and sugar are all that is needed to put up this delicious treat. These jams are never cooked, and because of that they are stored in the freezer, and can be kept frozen for up to a year. Once open, they should be used within a week. Care should also be taken when storing the jars of freezer jam in a chest type freezer to use freezer baskets, as the jars may break if anything heavy is tossed on top of them.

The following recipe is for strawberry freezer jam, however, other berries or even peeled, pitted peaches may be substituted for the strawberries.

Strawberry Freezer Jam

1 ½ quarts of strawberries
1 Tbsp lemon juice
5 cups sugar
1 package powdered pectin
1 cup cold water

Wash and remove caps from strawberries. Crush berries in large bowl. Measure 3 cups crushed berries. Add lemon juice and sugar, mixing well.

Let stand for 20 minutes, stirring occasionally. Dissolve powdered pectin in cold water in small saucepan. Bring the pectin solution to a boil over medium heat. Boil for 1 minute. Add the pectin solution to fruit mixture and stir briskly. Pour jam into clean, sterilized canning jars leaving ½ inch head space. Cover jars with lids and bands. Let the jars stand at room temperature for 24 hours, until the jam is set, then label and freeze. Makes seven ½ pint jars.

To thaw jam set it in the refrigerator overnight. If the jam separates upon thawing stir it up before serving.

Vegetables

Use garden fresh vegetables when they are at their best for eating. Sort vegetables for size, color and maturity. Discard all tough, over-ripe, or wilted produce. Set aside under-ripe produce for further ripening and later use.

Most vegetables need to be blanched, or scalded in rapidly boiling water, before they can be frozen. Blanching destroys the enzymes that cause undesirable changes in color, texture and flavor of vegetables.

Prepare the blanching kettle in advance. Fill a large kettle with water and bring it to a rapid boil over high heat. Use 1 gallon of water to every quart of vegetable being blanched.

After food is blanched it needs to be quickly cooled to prevent it from cooking completely. Fill a large pan or clean sink basin with ice-cold water. You will have to change the water frequently if you are blanching and freezing large quantities of vegetables.

Green peas

Fresh green peas are one of the easiest vegetables to freeze. Shell the peas, sorting out shriveled, immature or old peas as you shell. Place peas in a wire basket or small stainless steel strainer and immerse in rapidly boiling water. Scald peas for 1 ½ to 2 minutes. Plunge peas into ice-cold water, leaving them for 2 minutes. Drain peas and pack them quickly into freezer bags or waxed freezer boxes, label and freeze.

Green/Wax beans

Green and wax beans are frozen in a similar way. Leave long slender green or wax beans whole or cut or break them into desired length pieces. Blanche beans by putting them into a wire basket or strainer and immerse in boiling water for 3 minutes. Plunge the beans into ice-cold water for another 3 minutes. Drain beans and pack into freezer bags or waxed freezer boxes, label and freeze.

Sweet corn

Corn can be frozen on the cob, or cut from the cob. For corn on the cob, husk, de-silk, sort and wash corn. Blanche ears in rapidly boiling water for 7 to 11 minutes, depending on size of ears. Plunge ears into ice-cold water for 10 minutes to chill. Pack ears into freezer bags, label and freeze.

To freeze cut whole kernel corn, husk,

de-silk, sort and wash ears. Blanche corn in rapidly boiling water for 4 minutes. Plunge in ice-cold water for 4 minutes. With a sharp knife, cut kernels to a depth of 2/3 the depth of the kernels. It helps to have a clean wooden cutting board with a nail driven into the back side which protrudes into the front of the board. Stick the ear on the protruding nail to hold the ear still while cutting. Pack corn into freezer bags or waxed freezer boxes, label and freeze.

Greens

Greens of any kind can be frozen easily and with excellent results. Wash greens thoroughly, removing any tough stems, discolored or insect eaten leaves. Blanche greens in rapidly boiling water for 2 to 3 minutes, depending on age and size of leaves. Plunge greens into ice-cold water for 3 minutes to chill. Drain greens and pack into plastic freezer bags or waxed freezer boxes, label and freeze.

Meat and poultry

Meats bought from the supermarket should never be frozen in the containers in which they were bought. They should be repackaged in waxed freezer wrap, plastic freezer bags, or vacuum sealed plastic, then labeled and frozen.

The quality of frozen meat depends on the selection of good quality meat as well as the way in which the meat is handled during preparation for the freezer. The right wrapping materials are also very important in freezing meats and poultry. Poor wrapping technique will also lead to off flavors and freezer burn. Meats should be wrapped tightly, with no air pockets left in the food or wrap.

Only the freshest meats and poultry should be frozen for home use. As in any other food preparation, cleanliness is very important. Sanitize work surfaces and utensils before and after preparing meats for the freezer.

Freezer life of hams and bacon depends on the freshness of the meat, the curing process used, and the smoking. Salt has an undesirable effect on frozen meats. Usually hams and bacon will keep only from two to three months when frozen. The lower the salt content of the hams and bacon, the longer the holding time of the meats.

Meats

To freeze red meats, trim off excess fat, bone and gristle. Pack in plastic freezer bags or vacuum packs, or wrap in waxed freezer paper, label and freeze. Extra care should be taken when handling raw pork, chicken or rabbit. Wash hands immediately after handling these meats to prevent the spread of disease.

Poultry

To freeze poultry whole, remove any pin feathers or fine hairs. Rinse well and pat dry. Put fowl in large plastic freezer bag or vacuum bag, label and freeze. For fryer chicken, cut bird up into desired pieces and pack pieces into bags. If doing several birds at one time, pack all breasts or thighs in bags as desired, or mix pieces as needed. Backs and wings can be bagged separately for soups and stews. All should be labeled and frozen immediately after being packaged. Giblets should be packed as airtight as possible before being frozen. Giblets benefit greatly from being stored in vacuum sealed freezer bags.

Homemade pork sausage

An old New England recipe that is versatile and full of flavor.

20 lbs lean pork
10 lbs fat pork
1 Tbsp sugar
1 ½ tsp ground ginger
½ lb fine salt
2 Tbsp ground black pepper
3 Tbsp rubbed sage

1 Tbsp dried thyme leaves
1 tsp ground allspice
1 Tbsp crushed red pepper (optional)

Combine pork and grind medium fine. Add the seasonings to the meat and mix very well. Put the sausage mixture through the grinder using a fine blade setting. Package in plastic freezer bags, vacuum sealed containers or waxed freezer paper. Label, date and freeze.

For Italian sausage, add 1 Tbsp sweet basil to seasonings listed in above recipe. After second grinding, stir in 2 Tbsp fennel seed. Package, label and freeze.

Freezing eggs

What can you do with all those eggs your hens have blessed you with? You've eaten omelets, custards, and quiche until you can't stand the thought of any more. You know that winter is just around the corner when the hens slow down if not completely stop laying eggs. So how can you preserve the bountiful supply of eggs now so that you have a few put back for when the supply dwindles? Freeze them! Eggs freeze quite well if a few simple steps are followed. Remember, freezing eggs does not destroy any bacteria on them. Freezing merely slows the growth of bacteria to a halt, but that growth will begin once again when the eggs are thawed. Thawed frozen eggs should be treated as any opened, fresh eggs.

Prepare fresh eggs for freezing by sorting them, checking for any imperfections in the shell and discarding any that are cracked or chipped.

The easiest way to freeze eggs is in ice cube trays or muffin tins. Most ice cube trays' compartments will hold one large egg in each. Muffin tins will hold two to three eggs in each muffin cup, depending on size of muffin pan.

In a small bowl, break six eggs and add ½ tsp salt or 1 tsp sugar or honey to the egg and lightly stir together. The salt or sugar keeps the yolks from coagulating. Place 3 Tbsp of this mixture into each ice cube compartment, or 6 Tbsp into each muffin pan cup. Place the pans or trays in the freezer to freeze until firm. When frozen, remove from containers and put eggs in freezer bags, label and return to freezer. Be sure to label the bags with the number of eggs in each one. Eggs frozen in muffin tins are handy for recipes that call for 2 or 3 eggs, such as cakes, and cookies.

You can also freeze egg whites the same way, separating yolks from whites and stirring the egg whites up a little being careful not to introduce air into the whites. You don't have to add sugar or salt to egg whites as coagulation is not a problem. Place 2 Tbsp of egg white in each ice cube tray per one egg white. Freeze and package the same as whole eggs.

The yolks can also be frozen separately. Add the 1 tsp salt or 2 tsp sugar to 6 egg yolks to prevent coagulation. Put 1 Tbsp egg yolk in each ice cube tray compartment to equal one egg yolk in recipes.

When you want to thaw frozen eggs, remove them from the freezer and place them immediately in the refrigerator and let them thaw. Thawing may take several hours, so it is usually best to do this overnight for use the next day. Never thaw frozen eggs by letting them set at room temperature or by warming of any kind, as you may encourage the growth of salmonella. Never refreeze thawed eggs.

Homemade egg substitute

If you use this homemade substitute for scrambled eggs, cook it in a small amount of canola oil to keep them from sticking or being too dry. This is a great substitute for scrambled eggs and for use in muffins and other baked goods. It is not, however, a satisfactory substitute for use in custards and puddings.

1 tablespoon of nonfat dry milk powder
2 egg whites from large eggs

1 teaspoon canola oil
4 drops of yellow food color

Sprinkle powdered milk over egg whites, then beat them with fork or until smooth. Add canola oil and food coloring if desired, and beat until blended. This makes 1/4 cup, which is equal to 1 large egg.

Freezing fish and seafood

When properly handled, frozen seafood is nearly as flavorful and good in texture as if it were freshly caught. Fish and shellfish are extremely perishable, and need to be handled delicately, yet quickly upon catching or purchase.

Fish

All fish should be cleaned immediately and washed thoroughly in cold water. As in all other food preparation methods, cleanliness is a must.

Lean fish such as cod, perch, halibut and haddock and red or king salmon will all keep well frozen for up to six months.

Fatty fish like pink or chum salmon, mackerel, smelt, flounder, and tuna are more perishable and will keep well if frozen for not more than 3 months.

To prepare cleaned fish for freezing, scale fish if necessary, remove fins, tail, and head. Wash fish again in cold, running water.

You can package fish whole if you wish or you can fillet the fish first, or cut the fish into steaks or crosswise slices about 1-inch thick.

To fillet a fish, lay a larger fish on a clean cutting board. Run a thin-bladed sharp knife the entire length of the backbone, slightly above it to avoid cutting any bones into the flesh. Continue cutting along the side of the fish to separate the flesh from the backbone. Turn fish over, and repeat this process on the other side. Discard backbone, or use for making fish stock.

Lean fish can be dipped in a solution of 1 cup salt to 1 gallon cold water. Dip fish for 20 seconds before packaging. This helps to preserve flavor and texture of leaner fish. This step is optional but is a great aid in freezing lean fish.

Fattier fish should be dipped for 20 seconds in a solution of 2 tsp ascorbic acid in 1 qt of cold water. This dipping process will lessen the chance of rancidity and flavor change in fatty fish during storage.

Package fish in airtight freezer bags, or vacuum sealed containers. Label and freeze.

Crab and lobster meat

To prepare crustaceans for freezing, select active, live crabs or lobsters only. Avoid any that are inactive. Some people prefer to cook lobsters and crabs live, others prefer to butcher them first.

If you prefer to butcher them first, you can stun the crabs and lobsters first by plunging them into ice cold water for a few seconds, then remove legs and back. Remove innards. Rinse well under fresh, cold, running water.

Cook in boiling solution of 1 cup pickling salt and ¼ cup lemon juice added to each 1 gallon of fresh water. Drop crab or lobster pieces into this brine, and bring back to boiling. Boil hard for 15 minutes. Quickly remove the crab or lobster pieces and cool them quickly under cold running water until just cool enough to handle. Pick out the meat carefully from the shells. Remove any bits of tendon or shell that might stick to the meat.

Pack meat into containers leaving ½-inch of head space. Label and freeze.

Shrimp

Shrimp freezes best when packaged raw, but can also be cooked in table-ready dishes before freezing if desired.

To prepare shrimp for the freezer, sort

and wash shrimp, discarding any discolored shrimp. Cut off heads and devein. You can shell them if desired, or leave them in the shells. Wash the shrimp again in a mild salt solution of 1 tsp salt to each 1 qt cold water. Drain the shrimp well, pack into freezer bags leaving no head space. Seal tightly, label, and freeze.

Shellfish

Oysters, clams, and mussels and scallops are all extremely perishable and should be frozen within a few hours of being caught if at all possible. Oysters, clams and mussels should be frozen raw as they have a tendency to get tough in the freezer.

To prepare shellfish for the freezer, wash and sort them in cold water. Discard any opened or half-opened shellfish without using. Shuck then over a bowl to catch all their natural liquid. Reserve this liquid. To shuck them, or remove their shells, use a dull butter knife with a rounded tip, and insert this tip between the lips of the shell just beyond one end of the hinge. Twist to cut the muscle at this point, repeating with the other hinge. It will take some practice to become proficient at shucking shellfish.

Make a brine using 1 tsp salt mixed into 1 cup of water. Add the reserved liquid and mix well.

Pack shellfish meat in freezer containers and cover the meat with the brine liquid. Leave about 1-inch of head space. Seal, label, and freeze.

CHAPTER NINE
HOME CANNING

HOME CANNING IS ONE of the most rewarding and satisfying experiences you can have on the homestead. When you see the jars of home canned or bottled fruits and vegetables, soups and stews lining your pantry shelves, you know that good things await your family meals throughout the coming year.

You should start planning several weeks in advance of canning season. Decide on jar sizes and types. Larger families will want to put up produce in pints and quarts, while smaller families will most likely want to use half pint and pint size jars. Be sure to have enough jar rings and flat lids on hand. While you are in the middle of canning that bushel of tomatoes is definitely not the time to find out you did not buy enough supplies.

Inspect your pressure canner and

water-bath canner for any potential problems, making sure they are both clean inside and outside. Also examine your jars, caps and lids, discarding any that are chipped or dented.

Home Canning Equipment

Most of the equipment needed for canning can be found in the average homestead kitchen. If you plan on preserving any low-acid fruits or vegetables, or meats, poultry or fish, you will need to invest in a steam-pressure canner. Take time after purchasing your pressure canner to read the owner's manual and become familiar with your new canner.

- Jars sufficient for the task
- Metal bands and lids in amount equal to jars
- Jar labels
- Jar lifter
- Wide mouth funnel
- Ladle
- Steam-Pressure canner
- Water bath canner (large covered kettle with wire rack)
- Sieve or food mill
- Jelly bag or clean dish towel
- A food processor or food grinder
- Measuring cups
- Measuring spoons

- Sharp knife
- Colander
- Large saucepot or stock pot

There are only two methods of canning generally accepted as safe today, the steam-pressure method, and the water-bath method. In grandma's day folks canned and bottled home produce using methods like; oven canning, cold sealing, and open kettle method. These methods, although still used by some homemakers, are considered unreliable, and even dangerous. It is a far better practice to stick with methods that have been tested and proven to keep your food supply, and therefore, your family safe from food-borne illnesses.

Always begin by sterilizing your jars, lids, and bands, in boiling water in a water-bath kettle for 10 minutes. Turn off heat, leaving jars in water until ready to use.

Types of food spoilage

Bacteria, molds and yeasts are basic forms of plant life called micro-organisms. Their seeds are called spores, and they are found everywhere in our environment, the air, soil, water, even on our skin. These micro-organisms can be destroyed by proper cooking.

Botulism is a food poisoning caused by the toxin produced by the growth of spores from the Clostridium botulinum bacteria. This bacteria normally lives in soil, and the spores can survive in a dormant state until exposed to conditions that support their growth. They grow best in a low oxygen environment, which is why they grow so easily in incorrectly processed, sealed canning jars of low-acid foods. The spores are destroyed when these low-acid foods are properly processed in a steam-pressure canner.

Enzymes are natural substances found in all produce and meats. If not destroyed they will cause changes in the color, texture and flavor of your canned foods. Proper processing will destroy these enzymes.

Fermentation is caused by yeasts which have not been destroyed during processing. Some dishes are produced by the intentional fermentation of the food. What we are talking about is the unintentional fermentation of canned and bottled goods. As a general rule, fermented home-canned foods should not be used.

Flat-sour is the most common type of food spoilage in home-canned vegetables. It is caused by bacteria which give the food and unpleasant, sour flavor. As in other types of food spoilage, proper processing will destroy the bacteria that is responsible for flat-sour.

Molds can change the flavor and appearance of foods, and is caused by improper processing. It is best not to eat foods containing molds.

Steam-Pressure method

The steam-pressure method of canning uses a steam-pressure canner to process foods under pressure at a temperature of 240 degrees F, at up to 2000 feet above sea level. For altitudes up to 3000 feet above sea level, process at 11 ½ pounds pressure. Add another ½ pound pressure for every additional 1000 feet above sea level.

The steam-pressure method is recommended for processing low-acid foods such as green beans, corn, potatoes, and meats. The pressure canner is the only kitchen appliance that supplies a high enough temperature to kill the bacteria that causes botulism, as well as several other types of food spoilage.

There are two methods of processing vegetables using the steam-pressure method. They are; the hot pack method, and the cold or raw pack method. Not every vegetable can be processed safely and successfully using both methods.

Green or wax beans, carrots, corn, green peas, and asparagus all give good results with cold pack as well as hot pack methods. All other vegetables should be canned using the hot pack method unless otherwise stated. Try both methods to see which you prefer.

Green or wax beans

To can green or wax beans, pick the beans at the peak of freshness and when ripe, but not over-ripe. Thoroughly wash and sort beans, discarding any shriveled, insect eaten, or otherwise imperfect beans. Trim ends, and remove any strings. Snap or cut into desired size pieces, or leave whole.

For the hot pack method, put beans in stockpot, cover with boiling water and boil for 5 minutes, then drain. Fill sterilized jars one at a time with hot beans. Add 1 tsp salt per quart, 1/2 tsp per pint of beans. Cover beans with boiling water, leaving 1 inch head space.

To use the cold pack method, pack freshly snapped young beans tightly into sterilized jars, leaving 1 inch head space. Add 1 tsp salt per quart, ½ tsp salt per pint of beans. Cover beans with boiling water, leaving 1 inch head space per jar.

Wipe jar threads with clean damp cloth. Put jar lid on, rubber sealing compound down toward jar, and screw band on tight. Set aside until all jars are filled. Put jars into pressure canner containing an inch or two of hot water. Place canner over heat. Lock lid onto canner following manufacturer's instructions.

Leave vent open until steam escapes steadily for 10 minutes. Close vent. At altitudes of less than 2000 feet above sea level bring pressure to 10 pounds. Keep pressure steady for 20 minutes for pints, 25 minutes for quarts.

Remove canner from heat, allowing pressure to fall gradually to zero. Wait for 2 or 3 more minutes, then open vent slowly. Open canner and remove jars, putting them on a clean dry towel. Do not let sides of jars touch. Do not tighten bands. Let jars stand, out of drafts, for about 12 hours to completely cool.

Test the seals by pressing down on center of jar lid. If center stays down when pressed, jar is sealed. Remove bands for storage and store jars in a cool, dry and dark place.

Sweet corn

To can sweet corn, sort corn choosing only the freshest ears for canning. Sweet corn changes its sugar to starch very quickly after picking, which will greatly affect the quality of your finished product.

Husk the corn and remove the silk., and wash the ears. Cut kernels off ears with a sharp knife. A small block of wood with a nail driven into one side through to the other side works well to hold your ears stable as you cut the kernels. Do not scrape the ears after cutting.

For hot pack method, measure the kernels, and put into a saucepan add 1 tsp salt and 2 cups boiling water per quart of corn (½ tsp salt and 1 cup water per pint). Boil for 3 minutes. Pack boiling hot, into sterilized jars, leaving 1-inch head space.

For cold pack method, sort, husk, and wash ears. Cut corn from the cob. Pack corn loosely into sterilized ball jars, leaving 1-inch head space. Do not shake or pack down corn at all. Add 1 tsp salt per quart, ½ tsp salt per pint.

Wipe jar threads with clean, damp cloth. Adjust caps and process quarts 1 hour and 25 minutes, pints for 55 minutes, at 10 pounds pressure.

Green peas

Green peas are also easy to can or bottle at home. Sort and shell freshly gathered young green peas. Wash, and sort again, discarding shriveled, hard, or discolored peas. For hot pack method, boil small peas 3 minutes, boil larger ones for

5 minutes. Drain peas, reserving liquid, and pack loosely into sterilized jars, leaving 1-inch head space. Add 1 tsp salt per quart, ½ tsp per pint. Pour reserved liquid over peas in jars, leaving 1-inch head space.

For cold pack method, wash, drain and shell freshly gathered peas. Wash again. Pack loosely into sterilized jars, leaving 1-inch head space. Do not shake or pack down. Add 1 tsp salt per quart, ½ tsp salt per pint. Cover with boiling water, leaving 1-inch head space.

Adjust caps, process pints and quarts for 40 minutes at 10 pounds of pressure.

Greens

All greens, spinach, chard, mustard, turnip, kale, beet tops, poke, and dandelions are very good when canned. The instructions are the same for canning all greens. Wash greens thoroughly in several changes of water. Sort greens, discarding any large, tough stems, or insect eaten leaves. Leave greens whole or tear into pieces. Heat greens until wilted in just enough water to prevent sticking. To hasten wilting and prevent sticking stir a couple of times while heating. Pack hot into sterilized jars, leaving 1-inch head space. Add 1 tsp salt per quart, or ½ tsp salt per pint. Cover with boiling water, leaving 1-inch head space. Adjust caps, and process quarts 1 hour and 30 minutes, pints 1 hour and 10 minutes, at 10 pounds pressure.

Poultry

Duck, goose, chicken, quail, partridge, pheasant, and turkey all can or bottle well.

To can boned poultry, boil poultry until about 2/3 done. Remove skin and bones from meat. Reserve broth. Pack meat into hot, sterilized jars, leaving 1-inch head space. Add 1 tsp salt per quart ½ tsp salt per pint. Skim fat from broth, and reheat broth to boiling. Pour over meat in jars, leaving 1-inch head space. Adjust caps and process quarts 1 hour and 30 minutes, pints 1 hour and 15 minutes, at 10 pounds pressure.

To can bone in poultry, boil or bake cut up poultry pieces until about 2/3 done. Reserve broth. Pack into hot, sterilized jars, leaving 1-inch head space. Add 1 tsp salt per quart, ½ tsp salt per pint. Skim fat from broth, reheat broth to boiling. Pour over meat in jars, leaving 1-inch head space. Adjust caps and process quarts 1 hour and 15 minutes, pints 1 hour and 5 minutes, at 10 pounds pressure.

Meats

Pork, venison, veal, lamb, mutton, chevon and beef all can or bottle very well. Even tougher cuts come out of the jar tender and juicy.

Stew meat

Use beef or any other meat suitable for stewing. Trim meat of fat and gristle. Cut meat into 1 or 2 inch cubes. Simmer meat cubes until hot through in enough water to cover. Drain meat, reserving liquid. Season meat with salt to taste. Pack meat while hot into hot, sterilized jars, leaving 1-inch head space. Cover meat with the reserved liquid, leaving 1-inch head space. Adjust caps, process quarts 1 hour and 15 minutes, pints for 1 hour, at 10 pounds pressure.

Roasts

Trim excess fat and gristle off roast. Cut meat into 1 or 2 pound chunks. Roast meat until well browned, but not done, or brown roast in a small amount of fat. Salt to taste. Pack hot roast into hot, sterilized jars, leaving 1-inch head space. Cover meat in jars with hot gravy or broth, leaving 1-inch head space. Adjust caps, and process quarts 1 hour and 30 minutes, pints 1 hour and 15 minutes, at 10 pounds pressure.

Water-bath canner method

A water-bath canner is simply a covered Kettle of boiling water in which high acid foods are processed for a certain period of time. Processing is accomplished by boiling. Water-bath processing times in this book are given for altitudes of up to 1000 feet above sea level. For every 1000 feet higher above sea level you can adjust processing time simply by adding 1 minute to processing times stated for times up to 20 minutes, and adding 2 minutes for processing times stated for times over 20 minutes.

The water-bath canner method is used for high acid foods such as fruits and some vegetables. The water-bath method assures that all spoilage causing microorganisms are destroyed in these types of foods. Pickles, relishes, jams, jellies, fruit, sauerkraut, and tomatoes are all canned using the water-bath method.

In preparing many fruits for canning, you will want to use an ascorbic acid agent to prevent fruit from darkening. Commercially prepared mixtures, such as *fruit fresh*, can be purchased for this purpose. Ordinary vitamin C tablets may be crushed and dissolved in water, or you can use lemon juice or vinegar at the ratio of 2 Tbsp per gallon of water.

Many fruits are canned in sugar or honey syrups. You can make these syrups easily at home by mixing sugar and water or honey and water in certain ratios, depending on desired sweetness of syrup. For light sugar syrup use 2 cups granulated sugar to 4 cups of water. For medium sugar syrup use 3 cups of sugar to 4 cups of water, and for heavy sugar syrup use 4 ¾ cups sugar to 4 cups of water. Measure and mix sugar and water in a medium saucepan. Cook, stirring occasionally, until sugar dissolves. Keep syrup hot until needed, but do not let it boil down.

To make a medium honey syrup, combine 1 cup granulated sugar, 1 cup honey and 4 cups water in medium saucepan. Heat, stirring occasionally until sugar and honey dissolve completely. Keep syrup hot until needed, but do not let it boil down.

Peaches

To can peaches, sort, wash and drain only enough firmly ripe peaches for one canner load. Fill water-bath canner half full of warm water. Put canner on to heat. Prepare desired sugar or honey syrup.

Put the peaches in a wire basket or cheesecloth. Dip peaches into boiling water ½ to 1 minute to loosen skins.

Plunge peaches into ice cold water to prevent further cooking of peaches. Drain. Cut peaches in halves, peel, and pit. Drop peach halves into ascorbic acid and water solution.

You can also can peach slices by slicing the peach halves after they are peeled and pitted. Drop peach slices into ascorbic acid and water solution. Stand hot jar on clean dry towel on work surface. Pack peaches into hot sterilized jars, placing halves cavity side down, layers overlapping one another. Leave ½ -inch head space. Cover peaches with desired syrup, boiling hot, leaving ½ -inch head space.

Run a rubber spatula down the inside of jar, between fruit and jar side to release air bubbles. Add more syrup if needed.

Wipe jar threads with a clean, damp cloth. Put on lids and bands. Place jars in water-bath canner. Water in canner should be hot but not boiling. Add water, if necessary to bring water to cover jars by 1 or 2 inches. Place lid on canner, and bring water to a boil. Process pints 25 minutes, quarts 30 minutes at a steady boil. Remove jars from canner, placing them in a draft free place on a clean, dry towel for about 12 hours. Test for seal, remove bands, label, and store in a cool, dry, fairly dark room.

Pears

Put 2 inches of water in water-bath kettle and put on to heat. Make a sugar or honey syrup. Light or medium syrups work best for pears. Heavy syrups over-power the delicate pear flavor and are usually avoided.

Wash and sort pears, discarding any under-ripe or over-ripe fruit. Pare, core and cut pears into halves or quarters. Drop pears into water and ascorbic acid solution. Drain.

Cook pears in desired syrup for 5 to 6 minutes. Place hot jar on clean, dry towel on work surface to protect surface and prevent jar from slipping. Pack hot pears into hot sterilized jars, leaving ½-inch head space. Cover with boiling syrup, leaving ½-inch head space. Place caps and rings on jars, and place jars in wa-ter-bath kettle, and cover kettle. Process pints 20 minutes, quarts 25 minutes at a steady boil.

Remove jars, placing them on a clean, dry towel away from drafts to cool for 12 hours. Remove rings, test seals and label. Store in cool, dry, fairly dark room.

Tomatoes

Put 2 inches of water in bottom of water-bath kettle, and put on to heat. Wash and sort enough fresh, firmly ripe tomatoes for one canner load. Discard any tomatoes with decayed spots or cracks. Put tomatoes in a wire basket or cheesecloth. Place in boiling water for about ½ minute to loosen skins. Plunge tomatoes in ice cold water to prevent further cooking, and drain.

If desired for juice-pack, heat 1 quart of tomato juice in medium saucepan until hot but not boiling.

Cut out cores and trim away green spots. Peel tomatoes, cut in quarters or leave whole. Pack tomatoes into hot sterilized jars. Cover tomatoes with hot tomato juice, or press tomatoes down with a wooden spoon until juices fill spaces between the tomatoes. Add 1 tsp salt per quart, ½ tsp salt per pint. Run rubber spatula down inside of jar between tomatoes and jar side to release air bubbles.

Wipe jar threads, with a clean, damp towel, and put on lids and bands. Place jars in water-bath kettle of hot, but not boiling water. Add hot water until water covers jars by 1 or 2 inches. Place lid on kettle and bring water to boiling. Process pints 35 minutes, quarts 45 minutes, at a steady boil.

Remove jars from canner and place on a clean, dry towel, away from drafts to cool for 12 hours. Remove bands, test for seals, label and store in a cool, dry, fairly dark room.

Jams, jellies, and fruit butters

Jelly is made by cooking fruit juice with sugar, and often pectin. Jelly should be firm enough to hold its own shape when turned out from the jar, yet soft enough to be spread with a knife. Jellies can be transparent or translucent, depending upon the type of fruit used for the jelly as well as the method used to extract the juice.

Pectin is what makes jams and jellies gel. It is a natural ingredient derived from apples. Jams and jellies can be made without added pectin. Certain fruits such as apples, cranberries, quinces, gooseberries and red currants are high enough in natural pectin that added pectin is not necessary. All these fruits make very good jams and jellies whether added pectin is used or not. When making jam or jelly without added pectin, the juice and sugar mixture must be cooked for a longer period of time, so the process is a bit more time consuming.

Never make a double batch of any jam

or jelly if using added pectin. Results are far more satisfactory if a couple of smaller batches are made then by trying to make one large batch.

Apple jelly without added pectin

4 lbs tart red apples
4 cups water
3 cups granulated sugar

Use hard-ripe, tart fruit. Wash apples well, discarding stem and blossom ends. Cut apples into small chunks. Place apple chunks in large kettle. Add water to apples, and bring to a boil. Simmer 25 minutes.

Spread cheesecloth or place jelly bag in colander or strainer set inside a large bowl. Carefully dump apples and juice into the cloth or bag. Suspend the bag over the bowl, or, if using cheesecloth, bring corners of cloth toward center, forming a pouch. Suspend over the bowl. Allow juice to drain into bowl overnight.

In the morning, measure 4 cups of juice into kettle, and add sugar. Heat, stirring constantly until sugar dissolves to prevent scorching. Bring to a boil, and cook rapidly until jelly sheets off a spoon. Skim off foam, putting it into a custard cup or small jelly jar for immediate use.

Pour into hot sterilized pint, or half-pint sized jars, leaving ½-inch head space. Wipe jar threads with a clean, damp towel. Adjust caps, and place in water-bath canner containing about 3 inches of warm water. When all the jars are in the canner, add water until all jars are covered by 1 or 2 inches of water. Cover canner and bring to a boil. Boil half-pints and pints for 10 minutes.

Remove from canner, setting jars on clean towel away from drafts to cool for 12 hours. Test seals, remove bands, label and store in a cool, dry, fairly dark room.

Choke Cherry Jelly

A wonderful way to use the free fruits of nature!

3 lbs whole choke cherries or wild cherries.
½ cup water
4 cups sugar
1 pkg. powdered pectin

Wash, sort and stem fresh cherries. Do not crush chokecherries. Put whole cherries in kettle. Add water, and bring to a boil. Reduce heat and simmer 10 minutes. Extract juice by pouring fruit and juice through a jelly bag or 2 layers of cheese cloth draped over a colander or strainer that has been set into a large bowl. Suspend jelly bag or bring up corners of cheesecloth to make a bag. Let juice drain into bowl overnight.

Pour 3½ cups of juice into a large kettle, adding water if necessary to make 3 ½ cups. Stir pectin into juice. Bring to a boil over high heat, stirring constantly to prevent scorching. When boiling, add sugar all at once. Continue stirring and boil hard 1 minute. A tbsp of butter added will prevent foam from building up on the surface.

Remove from heat after 1 minute. Pour into hot sterilized jelly jars or pint canning jars, leaving ½-inch head space per jar. Wipe jar threads with a clean, damp towel. Put on lids and bands. Place jars in Kettle containing about 3 inches of warm water. When all jars are in canner, add water until all jars are covered by 1 or 2 inches of water. Cover canner and bring to a boil. Process jelly jars and pints in boiling water-bath canner for 10 minutes. Remove from canner, and set jars on a clean towel, away from drafts, to cool for 12 hours. Test seals, remove bands, label and store in a cool, dry, fairly dark room.

Two for one apple jelly and apple butter.

This is a very old recipe that makes twice the goodness out of one batch!

4 lbs fully ripe apples
6 lbs granulated sugar
1 bottle liquid pectin
Cinnamon
Allspice

Prepare fruit by washing and removing blossom and stem ends from apples. Cut into small pieces. Do not peel or core apples. Put apple pieces in large kettle. If apples are very soft or overly sweet, add 1 Tbsp lemon juice. Add 6 ½ cups water to apples in kettle, cover kettle and bring to a boil. Turn down heat and simmer for 15 minutes. Crush apples with potato masher and simmer, covered, for 5 minutes longer. Place apples in a colander or strainer that has been lined with a double thickness of cheesecloth and set into a large bowl or pot. Let apple juice drain. Do not squeeze juice out of fruit. Use 5 cups of apple juice to make jelly. Reserve fruit pulp for apple butter.

To make jelly, measure ½ bottle of liquid pectin into measuring cup. Mix 5 cups of juice with 7 ½ cups of sugar in a kettle. Bring to a boil over high heat, stirring constantly to prevent scorching. When boiling, add ½ bottle of pectin all at once. Bring back to a full rolling boil and boil hard for 1 minute, stirring constantly. Remove from heat, skim off foam with a metal spoon, and pour hot jelly into hot sterilized jelly jars or pint sized canning jars, leaving ½-inch head space per jar. Wipe jar threads with a clean, damp towel. Cover jars with lids and bands.

Place jars in kettle containing 2 or 3 inches of warm water. When all jars are in kettle, add enough water to bring water level 1 or 2 inches above jar lids. Cover kettle and bring to a boil. Process at a boil for 10 minutes for all sized jars. Remove jars from kettle, setting them on a clean, dry towel in a draft free location, and let cool for 12 hours. Test seals, remove bands and label jars. Store jelly in a cool, dry, fairly dark room.

To make apple butter, put remaining fruit pulp through food mill or sieve. Measure 5 cups of pulp and put in a kettle. Add 1 tsp cinnamon, and ½ tsp allspice stirring well to incorporate spices throughout pulp. Add 7 ½ cups sugar. Bring pulp to a boil, stirring constantly to prevent scorching. When boiling, pour in ½ bottle liquid pectin, all at once, continuing to stir constantly. Boil 1 minute longer. Remove from heat and pour into hot sterilized jelly or canning jars leaving ½ inch head space per jar. Wide mouth jars are ideal for apple butter. Wipe jar threads with clean, damp cloth. Put on lids and bands. Place jars in kettle containing about 3 inches of warm water. When all jars are in kettle, add enough water to cover jars 1 or 2 inches. Cover kettle, and bring to a boil. Process at boiling for 10 minutes for all jar sizes.

Remove jars from kettle, and cool on clean dry towel, away from drafts, for 12 hours. Test seals, label jars and remove bands. Store apple butter in a cool, dry, fairly dark room.

If you like a more robust flavored and darker colored apple butter, you can substitute 2 cups of brown sugar for 2 cups of granulated sugar in the above recipe.

Hot pepper jam

4 or 5 hot peppers, cored and cut into pieces
4 sweet green peppers, cored and cut into pieces
1 cup white vinegar (5 percent acidity)
5 cups sugar
1 pouch liquid pectin
Green food coloring (optional)

Note: Be sure to wear clean rubber gloves when working with hot peppers.

Put half the peppers and half the vinegar into blender container. Cover and process at liquefy until peppers are liquefied. Repeat with remaining peppers and vinegar. Combine peppers and vinegar mixture and sugar in a large saucepot and bring to a boil. Boil slowly for 10 minutes, stirring frequently. Remove from heat. Add liquid pectin all at once, and return to a boil. Boil hard for 1 minute. Skim off foam and add a few drops of green food coloring if desired. Pour hot jam immediately into hot, sterilized jelly jars or canning jars, leaving ½ -inch headspace. Wipe jar threads with a clean, damp towel. Place lids and bands on jars. Place jars in kettle containing about 3 inches of warm water. After all jars are in kettle, add water to cover jars 1 or 2 inches. Cover kettle and bring to a boil. Process by boiling for 10 minutes for all size jar.

Remove jars onto clean, dry towel and let cool in a draft free spot. Test seals, label, and remove bands. Store jam in a cool, dry, fairly dark room.

Grape jelly without added pectin

4 1/2 lb grapes
7 ½ cups sugar

Sort and wash just ripe concord grapes. Put the grapes into a large kettle and just cover with cold water. Simmer gently until grapes are soft, about 20 minutes. Line a colander or strainer with 2 layers of cheesecloth and set over a large bowl, or using a jelly bag, strain off the juice. Do not squeeze juice from grapes or your jelly will be cloudy. Measure 6 cups of grape juice into a large pot. Bring juice to a boil stirring frequently. Add 7 ½ cups of sugar to the boiling juice. Bring juice to a boil again, stirring frequently.

Cook until mixtures sheets from a metal spoon, or a drop of hot jelly on a plate sets up firmly.

Ladle into hot, sterilized jelly jars or pint canning jars, leaving ½ -inch head space. Wipe jar threads with a clean, damp cloth. Put lids and bands on jars and place jars in kettle containing about 3 inches of warm water. When all jars are in kettle, add enough water to cover jars 1 or 2 inches. Cover kettle and bring to a boil. Process by boiling for 10 minutes for all jar sizes.

Remove jars from kettle and set on clean, dry towel away from drafts to cool for 12 hours. Check seals and label jars. Remove bands and store jelly in a cool, dry, fairly dark room.

If spiced grape jelly is desired, a cinnamon stick and clove can be added to each kettle full of juice. Be sure to remove spices before bottling jelly.

Dandelion Jelly

1 quart packed yellow blossoms, all green parts removed
1 quart water
1 package powdered pectin
2 Tbsp Lemon Juice
4 ½ cups sugar
3–4 drops yellow food coloring

Place flowers in a large pot. Cover with cold water and bring to a boil. Boil the flowers for three minutes. Strain through cheesecloth or jelly bag.

Measure out 3 cups of dandelion juice in a large pot. Add the pectin and lemon juice. Bring juice mixture to a full rolling boil. Add sugar and food coloring. Return to a full boil and boil for 2 minutes. Pour into hot, sterilized jelly jars, leaving ½ -inch head space. Wipe threads with a clean, damp towel. Put on lids and bands. Place jars in kettle containing about 3 inches of warm water. When all jars are in the kettle add enough water to cover jars 1 or 2 inches. Process in boiling water for 10 minutes for all size jars.

Remove jars onto clean, dry towel and let cool, away from drafts, for 12 hours. Check seals, label, and remove bands.

Store in a cool, dry, fairly dark room.

Pickles and relishes

Pickles and relishes add spice to our most mundane meals. The perfect marriage of spices, sugar and vinegar blended with various fruits and vegetables creates a crisp, firm texture and a tangy, tart, or often sweet flavor.

Your corner supermarket will offer a wide variety of pickles and relishes to choose from, yet many homesteaders enjoy making their own pickles and relishes when their garden vegetables and fresh fruits are in season and in abundant supply.

Various types of pickled products can be made depending on the ingredients used and the methods of preparation.

Brined pickles or fermented pickles go through a curing process in a brine (salt and water) solution for one or more weeks. Curing changes the color, flavor and texture of the pickles or relish being cured. If the product is a fermented one, the lactic acid produced during fermentation helps preserve the product. In brined products that are cured but not fermented, acid in the form of vinegar is added later to preserve the food.

Fresh pack or quick process pickles are covered with a boiling hot mixture of vinegar, spices and seasonings. Sometimes, the product may be brined for several hours and then drained before being covered with the pickling liquid. These pickles are easy to prepare and have a fresh, tart flavor. Fresh pack or quick pickles have a better flavor if allowed to stand for several weeks after they are sealed up in jars.

Fruit pickles: These pickles are prepared from whole or sliced fruits that are simmered in a spicy, sweet-sour syrup made with vinegar or lemon juice and usually, sugar or honey.

Relishes: These are made from chopped fruits and vegetables cooked to desired consistency in a spicy vinegar and sugar solution.

The level of acidity in pickles and relishes is as important to its safety as it is to its taste and texture. Never change the proportions of vinegar, food or water in a recipe.

Pure granulated salt, such as "pickling" or "canning" salt is the only salt that should be used. It can be purchased from grocery, hardware or farm supply stores. Other salts contain anti-caking materials, and iodine, all of which may make the brine cloudy. Never alter the salt concentrations in fermented pickles. Proper fermentation depends on correct proportions of salt and other ingredients.

Use cider or white vinegar of 5-percent acidity (50 grain) only. This is the range of acidity for most commercially available vinegars. Cider vinegar has a good flavor and aroma, but may darken white or light-colored fruits and vegetables. White distilled vinegar is often used for light or white pickles such as onions, cauliflower and pears where clearness of color is desired. Never use homemade vinegar or vinegar of unknown acidity in pickling. Do not dilute the vinegar unless the recipe specifies. If a less sour product is preferred, add sugar rather than diluting the vinegar.

Use only white granulated sugar unless the recipe calls for brown sugar. White sugar gives a product a lighter color, but brown sugar may be preferred because of the more robust flavor it imparts. Sugar substitutes are not usually recommended, as heat and/or storage may alter their flavor. Sugar also helps to plump up the pickles and keep them firm.

Use fresh whole spices for the best quality and flavor in recipes that call for whole spices If you substitute ground spices when the recipe calls for whole spices you may cause the pickles to darken and become cloudy. Pickles will also darken less if you tie the whole spices loosely in a clean white cloth or cheesecloth bag and then remove the bag from the pickles before filling the jars. Relishes and a few pickle recipes will sometimes call for ground spices. Use only the freshest ground spices in these recipes as well. Ground spices are always mixed directly into the vinegar and sugar solution. Spices deteriorate over time and will quickly lose their pungency if stored in the heat and humidity. Store any unused spices in an airtight container in a cool place.

When brining pickles, hard water may interfere with the formation of acid and prevent the pickles from curing properly. To soften hard water, simply boil it for 15 minutes and then let it set, covered, for 24 hours. Skim any scum that appears on the water's surface. Slowly pour the water out of the containers so that the sediment at the bottom of the container will not be disturbed. Discard the sediment. The water is now ready for use. Distilled water can also be used in pickle making, but is a bit more expensive.

If good-quality ingredients are used and up-to-date methods are followed, the lime and alum are not needed for crisp pickles. Soaking cucumbers in ice water for four to five hours prior to pickling is a safer method for making crisp pickles.

Alum may be safely used to firm fermented cucumbers, but does not work with quick process pickles.

The calcium in lime does work to improve the firmness of pickles. If you choose to use lime, purchase food-grade pickling lime from your supermarket. Do not use agricultural or burnt lime. Food-grade lime may be used as a lime-water solution for soaking fresh cucumbers 12 to 24 hours before pickling them. However, and this is very important: *All excess lime absorbed by the cucumbers MUST be removed in order to make the pickles safe to eat.* To remove excess lime, drain the lime-water solution, rinse and then re-soak the cucumbers in fresh water for one hour. Repeat this rinsing and soaking procedure twice more.

Equipment

The right equipment helps prevents pickle failure and saves time and energy.

Pickles can be fermented in large stoneware crocks, large glass jars or food-grade plastic containers. To determine if a plastic container is food-grade, check the label or contact its manufacturer, or line the questionable container with several thicknesses of food-grade plastic bags. Do not use aluminum, copper, brass, galvanized or iron containers for fermenting pickles or sauerkraut. The container needs to be large enough to allow several inches of head space between the top of the food and the top of the container. Usually a 1-gallon container is needed for each 5 pounds of fresh vegetables.

After the vegetables are placed in the container they are covered with brine, and they must be *completely* submerged in the brine. A heavy plate or glass lid that fits down inside the container can be used. If extra weight is needed, a covered canning jar filled with water can be set on top of the plate or lid. You may have to use a bigger jar with more water inside if you need more weight to keep the vegetables completely submerged. The vegetables should be covered by 1 to 2 inches of brine. Another option for submerging the vegetables in brine is to place one food-grade plastic bag inside another and fill the inside bag with some of the pickling brine, in case the bags are accidentally punctured. Freezer bags sold for packaging turkeys are the right size for 5-gallon containers. Seal the end securely. Then use this filled bag as the weight on top of the vegetables.

For fresh pack pickles, the pickling liquids should be heated in a stainless steel, aluminum, glass or enamelware saucepan. Do not use chipped enamelware, copper, brass, galvanized or iron utensils. These metals can react with acids or salts and cause undesirable color changes in the pickles, as well as further pitting and deterioration of your equipment.

For short-term brining or soaking, use crocks, saucepans or bowls made from stoneware, glass, stainless steel, aluminum or enamelware. Except for the aluminum, the same containers can be used for soaking vegetables in lime. Lime pits aluminum containers and can cause an increased level of aluminum in your finished pickles.

Not all pickles have to be canned or bottled. A few, like the bread and butter pickle recipe given below, call for the pickles to be refrigerated instead, and the pickles can be held for several months this way without any further processing.

Refrigerator bread and butter pickles

This is a large recipe for spicy-sweet pickles that store in the refrigerator. If the 8 quarts this recipe makes are too many pickles for you to store, you can cut the recipe in half.

25 cucumbers
6 large onions, thinly sliced

2 green bell peppers, diced
3 cloves garlic, chopped
½ cup pickling salt
3 cups cider vinegar
5 cups white sugar
2 Tbsp mustard seed
1 ½ tsp celery seed
½ tsp whole cloves
1 Tbsp ground turmeric

Wash and sort cucumbers, remove blossom ends. Slice cucumbers thinly.

In a large bowl, mix together cucumber slices, onion slices, diced peppers, chopped garlic and salt. Allow this mixture to stand for 3 hours.

In a large saucepan, mix the cider vinegar, white sugar, and all spices together. Bring to a boil.

Drain liquid from the cucumber mixture. Rinse, and drain again. Stir the vegetables into the boiling vinegar solution, leaving on heat. Remove from heat shortly before the pickles return to a boil. Pack hot into hot, sterilized canning jars. Wipe threads with a clean, damp towel. Put on lids and screw bands. Store pickles in the refrigerator until eaten. Pickles can be stored for several months.

Sweet zucchini relish

10 cups ground zucchini
4 cups ground sweet onion
2 large ground, red bell peppers
2 large, ground, yellow bell peppers
1/3 cup pickling salt
4 ½ cups sugar
½ tsp turmeric
1 Tbsp celery seed
1 Tbsp dry mustard
1 tsp ground black pepper
3 cups vinegar, white or cider

Mix Zucchini, onion and peppers well. Place mixture in a very large plastic or glass bowl. Add pickling salt and mix well to distribute salt throughout vegetables. Let stand covered, over night.

Next morning, drain and rinse vegetables in cold water. Drain again. Place vegetables in a large stock pot or kettle. Add all other ingredients to pot and mix well. Simmer relish for 20 minutes. Pour into hot, sterilized canning jars, leaving ½-inch head space. Wipe jar threads with a clean, damp towel. Put on lids and bands. Place jars in kettle filled with 3 inches of warm water. When all jars have been added to kettle, fill kettle with warm water until jars are covered with 1 or 2 inches of water. Process at boiling for 15 minutes for all jar sizes.

Remove jars and set on clean, dry towel in a draft-free spot to cool for 12 hours. Test seals, label and remove screw bands. Store relish in a cool, dry, fairly dark room.

Old time garlic dill pickles

10 lbs pickling cucumbers (3 ½ to 5 inches in length)
¼ cup whole mixed pickling spices
2 bunches of fresh dill
2 cup white vinegar
1 1/2 cup coarse pickling salt
2 gallon of fresh water
10 garlic cloves, peeled

Sort cucumbers, removing any that are bruised, over-ripe or damaged in any way. Scrub cucumbers with a soft vegetable brush. Cut 1/16 inch the blossom ends.

Place half the pickling spices and 1 bunch of dill in the bottom of a 5 gallon plastic bucket (food grade only). Mix vinegar, pickling salt and water, stirring to dissolve the salt. Pour ½ of this mixture over cucumbers, reserving the remainder. Add garlic, remaining spices and dill to cucumbers in plastic bucket.

Select a glass or ceramic plate that is smaller than the opening at the top of the plastic bucket.

Using remaining brine, fill a zipper lock plastic freezer bag with brine. Make sure bag is tightly sealed, and place it on the plate. Place plate and bag atop pickles in bucket to submerge pickles completely in brine. If pickles don't submerge completely add more of the remaining brine to pickles in bucket. Then, replace plate and bag. Cover the bucket with a large, clean towel, and put bucket aside to work in a cool spot at room temperature (70 to 75 degrees is ideal)

Check the pickles every day and skim off the film that develops on the surface of the brine. It is important that you are diligent in removing this scum, as it is caused by a yeast that will cause your pickles to spoil if not removed. This scum usually starts to develop after only a day or two. Make sure pickles are still completely submerged in brine. You can make more brine at the original proportions if you need to add more brine to your pickles.

Let the cucumbers ferment in the bucket until they are evenly colored olive green or until they are translucent throughout. This should take 2 ½ to 3 weeks. Don't taste the pickles until this color has been reached. At the end of 2 ½ to 3 weeks it will be safe to taste your pickles. Do not leave your pickles to ferment in the bucket for more than 3 weeks.

Strain the brine off the pickles through cheesecloth, catching the brine in a large bowl. Pour brine into a stainless steel or enamel pot, and bring to a boil. When boiling turn heat down to simmer. Pack pickles into hot, sterilized canning jars leaving ¾-inch head space. Add a couple of sprigs of dill to each jar. Pour hot brine over pickles in jars leaving ½-inch head space per jar. Wipe jar threads with a clean, damp cloth. Apply lids and bands.

Place jars in a water-bath canner that has 2 to 3 inches of warm water in the bottom. When all jars are in the canner add more warm water until all jars are covered by 1 or 2 inches. Process by boiling for 15 minutes for all jar sizes.

Remove jars from canner and set on a clean, dry towel in a draft-free place too cool for 12 hours. Test seals, label and remove screw bands from jars. Store pickles in a cool, dry, fairly dark room for at least 2 weeks before tasting to mellow flavors.

Pickled beets

10 lbs fresh beets
2 cups white sugar
1 Tbsp pickling salt
1 quart white vinegar
¼ cup whole cloves

Wash and sort beets. Place beets in a large stockpot with water to cover. Bring to a boil, and cook until tender, about 15 minutes, depending on size of beets. Drain, reserving 2 cups of beet water, cool and peel beets. If beets are large, you can either slice them, or cut them into quarters. Leave small beets whole.

Pack beets into jars leaving ¾-inch head space per jar. Add several whole cloves to each jar.

In a large saucepan, combine sugar, beet water, vinegar, and salt. Bring to a rapid boil. Pour hot brine over beets in jars leaving ½-inch head space.

Wipe jar threads with a clean, damp towel. Place lids and bands on jars. Place jars in kettle containing 2 or 3 inches of warm water. When all jars are in kettle, bring water level up to 1 or 2 inches above tops of jars. Process by boiling for 15 minutes, for all jar sizes.

Remove jars to clean, dry towel, away from drafts, to cool for 12 hours. Test jar seals, label jars, and remove bands for storage. Store pickles in cool, dry, fairly dark place.

Spicy tomato salsa

You can adjust the flavor and heat of this salsa by simply adding more or less hot peppers or cilantro in this recipe.

8 cups peeled, cored, chopped fresh tomatoes
2 jalapeno peppers with seeds, chopped
2 cups chopped onion
8 cloves garlic, minced
¼ cup chopped fresh cilantro
1 Tbsp pickling salt
¾ cup cider vinegar
¼ cup lime or lemon juice

Combine all ingredients in a large pot or kettle. Bring to a boil over medium high heat. Reduce heat and simmer for 15 minutes. Pour hot salsa into hot pint jars, leaving ¼-inch headspace. Put on jar lids and screw bands. Place jars in kettle that contains about 3 inches of warm water. When all jars are in kettle add enough warm water to cover jars 1 or 2 inches. Process in a boiling water bath for 15 minutes, for all jar sizes.

Remove from kettle and set jars on a clean, dry towel away from drafts to cool for 12 hours. Test jar seals, label jars and remove screw bands. Store salsa in a cool, dry, fairly dark room.

CHAPTER TEN
Dehydrated Foods

Most of us homesteaders prefer to preserve our garden bounty by freezing or canning for future use. More and more homesteaders, however, are turning to dehydrating as a valuable way to preserve food. Dehydration is the process of removing or reducing the moisture content of foods. Dehydration is also called drying.

Drying foods is quite inexpensive and very easy to do. From a nutritional standpoint, drying compares well with other methods of food preservation.

On average, the nutritional value retained in dried foods is nearly equal to that of frozen foods, and often exceeds that of canned goods. Dried foods retain much more nutritional value because they have not been subject to the prolonged high temperatures required in canning or bottling. Dried foods can also be stored on the pantry shelf rather than in the freezer or refrigerator. They are lightweight and therefore easy to take camping or on hiking or biking trips.

Dehydrating fruits and vegetables concentrates the natural sugars found in these foods, making them sweeter and more flavorful.

Most foods dry well with very little pre-treatment. An ascorbic acid product of some sort will be needed for fruits like apples and peaches to prevent them from turning an unappealing brown color. This can be purchased in a commercial, ready to use powder, or can be obtained by crushing vitamin C tablets, or using powdered vitamin C, available through some drug stores. Ordinary vinegar or lemon juice can be used to prevent browning in susceptible fruits as well.

You can purchase many different kinds of commercially available dehydrators. These usually consist of several trays stacked on top of one another above a fan-driven motor assembly. The motor sends heated air up through the stacked trays slowly dehydrating your food. There are still-air dehydrators available, but these often give unsatisfactory results and can lead to food spoilage. Stick with the forced air models if possible.

Kitchen range dehydrating

In a pinch you can use your stove's oven for a dehydrator. This is too expensive to be a viable option for frequent use, but would work for those occasional food drying jobs.

Most unsulfured fruits, fruit leathers, vegetables, meats, fish, nuts and cereals can be successfully oven dried. Sulfured fruits should never be oven dried because the sulfur dioxide fumes can cause irritation to anyone in the vicinity of the kitchen. Any other method of pretreatment will be completely compatible with oven drying.

Convert your oven racks to drying racks by stretching muslin or cheesecloth across the oven racks. Secure with toothpicks, straight pins or long running stitches. Alternate trays in the oven periodically to assure even drying. Set oven control at its lowest setting, but not below 140 degrees F. Most fruits and vegetable dry well at 150 degrees F. Meats should be dried at 150 to 160 degrees F. It is best to use a reliable oven thermometer while oven drying. If you are using an electric oven, wedge a pot holder between oven and door to allow a 1" opening to ensure good air exchange. Moisture from the drying food will vent through this opening. The door on a gas oven can be kept closed because the moisture will vent via the exhaust flue.

Solar dehydrating

Solar drying, or sun-drying is gaining in popularity once again. For centuries our ancestors used the sun to dry their winter stores of foods. The flavor of sun-dried tomatoes and salt-fish is legendary. The methods used in olden days to sun-dry these foods are every bit as effective and efficient today. However, solar drying is not an effective method of drying foods in all parts of the country. In the deep south and northeast the humidity level is usually too high to obtain a desirable or even safe level of dehydration, despite the heat of the day. The frequent afternoon thunderstorms in these regions also make solar drying difficult at best. The key requirements for successful solar-drying is; continued sunshine, high temperature, and low humidity. Of course you would also want to avoid sun-drying foods if you live in an area that is often smoggy or if you live along a frequently used highway.

You can purchase or make solar dehydrators, but that is unnecessary as you can successfully dehydrate foods very nearly for free by following a few basic instructions which I refer to as PAPS.

Prepare; prepare foods for drying, including any pretreatment desired. **Allow**; allow the sun time to work, often several days. Fruit leathers can be sun dried very successfully in one day if the sun is hot and the humidity low. **Pasteurize**; pasteurize your food in the oven to kill harmful organisms such as insects and their eggs. **Store**: store dried foods properly in airtight containers, and only for duration recommended for that particular food.

Prepared foods are placed on drying trays. Today we can use stainless steel screening and thin wood strips as materials for our home-constructed drying trays. Aluminum screening reacts with the acids in various fruits, so it is less desirable. Do not use galvanized metal, copper, fiberglass or vinyl screening for drying foods unless they are well covered with two layers of cheesecloth.

Select a warm, sunny day when the humidity is lowest. You will not want your foods to be in extreme direct sun for the full course of the day. Continuous full sun often results in a product that is *case hardened,* or drying to a hard crust on the outside while the inside is still moist. This condition can be likened to sunburn or sun scorching.

Place trays of food away from dusty

roads and yards. The shelf behind the rear seat of a vehicle makes a convenient solar dehydrator. If solar drying your foods outdoors, in your yard, elevate the trays at least one inch above the table using bricks to allow good air circulation beneath the food. Cover the food being dried with a muslin or cheesecloth tent to protect it from insects.

Allow the heat of the sun to do its magic on your foods, checking every now and then to be sure things are going smoothly. Move the racks out of the direct sun as needed to prevent case hardening or scorching.

Bring your foods indoors at dusk to prevent condensation from forming on the food, and to discourage night-time scavengers such as raccoon from helping themselves to your product. Depending upon the food being dried in this manner, it may take several days to achieve a sufficiently dried product.

Pasteurize your sun-dried product to destroy any insects or their eggs that may be on them, and to remove additional moisture in the thicker pieces. To pasteurize dried foods, heat them in a 150 degree F oven for 30 minutes.

The drawbacks to solar drying are that many foods loose or change color from exposure to the sun. You can dry your foods in the shade on a hot day, to help preserve food color, but this is not technically considered solar drying. As usual in all food preservation practices, you can pre-treat foods with an ascorbic acid solution to prevent browning.

Fruits that are most suitable for solar-drying are; apples, apricots, cherries, dates, figs, nectarines, peaches, pears, and plums.

Vegetables that are most suitable for solar-drying are; mature shell beans, peas, soybeans, hot peppers, sweet corn, sweet potatoes, and shredded mixed vegetables.

Most other fruits and vegetables are really not well suited for solar-drying. Herbs are generally not recommended for solar drying because the action of the sun on these plants causes them to lose color, flavor and aroma.

Electric dehydrator

Electric dehydrators are the most commonly used dehydrators because of their convenience. You don't have to wait for a sunny day to dry your foods. They are also fairly inexpensive to buy. You can also build your own dehydrator for very little money. Plans for building an electrical dehydrator are included in this book.

Preparing foods for drying.

To prepare foods for drying, sort and wash desired produce. Peel and core, then chop, slice or shred food in desired manner. You can pre-treat vegetables by steam blanching them to stop the bacterial and enzymatic actions that will lead to decay. Steam blanching of vegetables also protects some of the nutrients and loosens the tissues so that drying is actually faster. Steam blanching is not a required step, and some vegetables actually dry very well without any pretreatment, such as zucchini. Potatoes, however, need to be blanched to prevent them from turning an unappealing black color when dried.

There are several treatments available to use on fruits that are to be dried. The anti-oxidants such as ascorbic acid solutions and powders can be used to prevent browning of oxidizing susceptible foods such as apples, peaches, and apricots.

Syrup blanching fruits

Fruits can be blanched in heavy syrup. This process adds somewhat to the drying time of the fruit, and requires a little more diligence on your part to be sure your fruit doesn't stick to the drying trays, but the sweet confection created by this method is loved by many homesteaders. Unfortunately, insects love it too, and so when drying syrup blanched fruits outside in the sun, extra caution should be used to protect the fruits from insects.

To syrup blanch fruit, prepare fruits by washing and sorting, peeling and slicing as desired. Hold cut fruit in an ascorbic acid solution until desired quantity of fruit is prepared for syrup blanching.

Prepare a sugar syrup by mixing 1 cup sugar, 1 cup white corn syrup and 2 cups water. Bring this mixture to a boil. Add 1 ½ pounds of prepared fruit. Simmer for 10 minutes. Remove fruit from heat and leave it in the hot syrup for 45 minutes. Drain fruit, and rinse lightly in cold water. Place fruit on dryer trays for drying.

Checking fruits

Many fruits, such as cherries, bramble fruits, blueberries, plums, and grapes have a wax-like coating on them that prevents the fruits from drying out naturally. This water-proof waxy coating should be removed prior to drying to achieve the best results. Along with removing this coating, you will also be *checking* the skins of these fruits. Checking means to make minute cracks along the surface of the fruits. This allows for better drying and less likelihood of case-hardening. Checking is done by dipping fruits quickly in boiling water, just long enough to crack the skins.

Checked fruits are slightly less flavorful then fruits pretreated by other means.

Steam blanching vegetables

As stated earlier, steam blanching is not mandatory for drying vegetables, but it does help some vegetables to retain their flavor, color, and texture.

To steam blanch vegetables, put several inches of water in a large kettle with a close-fitting lid. Heat the water to boiling, and set a wire basket or metal colander over the boiling water so that the basket does not touch the boiling water. Place your prepared vegetables in the basket making sure the food is no more than two inches deep. Cover the kettle and let the food steam for the specified time, checking halfway through to be sure all parts of the food are getting equally steamed. Remove the food from the steamer and spread it out on paper towels or clean cloth towels to remove excess moisture. Do not try to squeeze moisture out of the produce. Lay the steamed foods on the drying trays allowing nothing to overlap.

Sulfur and Lye

For years both sulfur and lye were used to pre-treat fruits and vegetables for drying. Lye use has been discontinued in recent years as a food pre-treatment because of the fact that lye is a poison and unless used in exact quantities, and very controlled methods, the opportunity for accidental poisoning and chemical burns is ever present.

Sulfur is still used on occasion for a few foods, but is not a required step for any foods. Using sulfur to pre-treat foods can be irritating to eyes and breathing passages, and should be done out of doors in the fresh air. Sulfur is not toxic to humans when used to pre-treat foods for drying, and does help to keep certain vitamins in the foods that might otherwise be lost during the drying process.

Use sublimed sulfur for food purposes. It is a yellow powder that has no taste and a faint odor. You can get sublimed sulfur from your druggist or order it online.

You'll also need to supply a large box, such as appliances come shipped in, to use as a sulfuring box. This box should be large enough to accommodate 6 trays high and have at least 6 inches to spare in front of the trays to allow room for the sulfuring dish. At the bottom of the box, in one corner, cut out a slot 6 inches long by 1 inch high. This is the air intake that will keep the sulfur burning. At the top of the box, opposite the side the intake slot is, cut a small hole about the diameter of a pencil. Both of these holes will be covered after the sulfur has burned out completely.

You'll have to weigh your prepared fruit to keep the amount of fruit and sulfur balanced. Stack your trays of weighed, prepared fruit on top of one another with 1x2 strips of lumber between each tray to separate them and allow for free air flow between trays. Place the bottom tray up on bricks to raise it at least 4-inches above the ground or table.

Place your prepared box over the trays of fruit. Place sulfur powder in a shallow crockery dish at the rate of 1 level teaspoon for each pound of prepared fruit. Smooth the sulfur so that it does not exceed ½ inch deep in the dish. Light the sulfur and place the dish under the box next to the drying trays. Let the sulfur burn for 15 minutes or until completely burned. Check on the sulfur periodically to be sure the flame has not gone out before the sulfur is completely burned.

After the sulfur has finished burning, tightly close off the intake slot and the small hole at the top of the box. Leave the box with fruit trays beneath it for required length of time. Slices of apples, apricots, peaches and pears that are to be dried in an electric dehydrator should be sulfured for about 20 to 30 minutes. Apricot, peach and pear halves should be sulfured for 45 minutes to an hour, depending on thickness of halves. If you are planning on solar drying these foods, sulfuring times should be doubled.

After the required time has elapsed, remove box from atop the trays, pulling the top of the box toward your body to allow the remaining fumes to vent away from you. Remove trays, one at a time, being careful not to spill any juice that may have collected in the hollows of the fruits.

Continue to dry in electric dryer or out in the sun, as desired.

Do not store sulfur treated dried foods in any metal container, unless food is sealed in a plastic bag first. If using a glass or plastic container with a metal lid, first line the lid with plastic wrap or waxed paper. Sulfur fumes still on the foods will react with unprotected metal. Glass or plastic storage containers are the best choice for storing foods treated with sulfur.

General drying techniques for oven or dehydrator

Arrange the prepared food on the drying trays in a single layer. Try to prevent overlapping pieces. Most types of fruits and vegetables can be dried together, although you'll want to avoid drying strong smelling vegetables, such as onions or garlic, with other produce.

Preheat electric dehydrator for several minutes. Load trays into electric dehydrator, and follow manufacturers operating instructions. Rotate trays periodically. Remove food as it becomes dry. Food on the edge of trays often dries faster than that in the center. Trays on the bottom and top of dehydrator will also often dry faster than the trays in the middle. Be diligent in your attention to nearly dried foods to prevent over-drying.

If using your kitchen range's oven for drying you'll want to preheat the oven to 160 degrees F. The temperature of the oven will drop about 20 degrees when you put the trays of cool, moist food in it. Keep the room well ventilated while the oven is going to allow for faster drying. A small fan set outside the oven, off to one side, will also aid in speeding drying time and will result in a more satisfactory product. Rotate trays to maintain even drying. Remove food as it dries. Be sure to prevent over-drying of foods.

Conditioning dried foods

After drying is completed, loosely pack food in plastic or glass containers until about 2/3 full. Metal containers such as cans may impart an unpleasant flavor to dried produce. Cover containers tightly and let them stand for 3 or 4 days. Any excess moisture in some larger pieces will be absorbed by the drier pieces. Shake the containers daily to keep pieces from sticking together, and to check for any condensation within the containers. Correctly dried foods will not produce any condensation. If condensation does occur, the fruit should be returned to the drier for further drying. If mold or other spoilage is noted, discard food and clean the container with hot soapy water.

Pasteurizing

Washing and retreating foods does not always remove insect larvae, so fruits and vegetables that have been exposed to insects before or during the drying process, namely solar dried foods, or foods dried in an electric dehydrator out of doors, must be pasteurized to prevent spoiling during storage. There are two methods of pasteurization that can be used with equal success.

The first method is the freezer method. After the fruit has been dried and conditioned, seal it up in heavy plastic bags. Place the bags in a freezer set below 0 degrees F (-20 C) for at least 48 hours or up to 2 weeks before storing.

The second method is the oven method. This method will result in somewhat higher vitamin loss in the dried product. After the food has been dried and conditioned, layer the dried food loosely on a jelly roll or roasting pan. Place the pan in an oven that has been preheated to 175 degrees F (70 degrees C) for 15 minutes. Remove the food, cool completely and package for storage.

Packaging dried foods for storage

Dried foods should be cooled completely and conditioned before packaging. If dried out of doors in any manner, dried foods should be pasteurized before packaging. Every time dried foods are exposed to air they absorb moisture, so only package foods in small amounts that can be used within a few days of opening.

Vacuum sealed bags, or heavy zipper lock plastic storage or freezer bags are ideal for storing dried foods. Lightly pack dried food in bags and either vacuum out the air or squeeze out as much air as possible before sealing the bags.

Label the bags and store them inside large plastic tubs, glass jars, or metal containers. Silica gel packets placed inside these larger storage containers will help keep the moisture content inside to a minimum, which will ensure longer storage life.

Apples

Choose apples that are mature but firm, and free from bruises and insect spots. Wash, and peel if desired. Core and slice apples into ¼-inch slices. Hold in solution of 1 teaspoon ascorbic acid per quart of water to pre-treat.

Sulfur ¼-inch slices for 45 minutes to 1 hour. Spread slices on dryer trays and process until apple slices are pliable to crisp in texture. Apples have a longer storage life if they are dried to slightly crisp. Store in airtight, labeled storage bags.

Dried apples make a delicious and wholesome snack. Re-hydrated dried apples make wonderful applesauce, apple pies, cobblers, and crisps, and apple cakes. Re-hydrated apple rings can be fried to make luscious fried apples and apple dumplings.

Apricots

Select apricots for drying when they are at the peak of ripeness. They should be plump, but not too soft. Sort, wash and halve apricots, remove pits, and hold fruit in an ascorbic acid solution using 1 tsp ascorbic acid to each quart of water used. When you have a whole batch ready to process you can prepare your sulfuring box and sulfur the apricots skin side down for 2 to 3 hours. If possible, expose sulfured apricots to the heat of the sun for 6 hours to deepen their color and improve the flavor.

If you prefer not to sulfur your fruits you can pre-treat them in a syrup blanch. The appearance of syrup blanched apricots is not as attractive as is the sulfured fruits, and the nutritive value of the fruits is less as well.

Spread apricots on dryer trays in single layer and process until fruits are dry yet pliable with no pockets of moisture present.

Store in airtight plastic bags, and label.

Bananas

Wash and sort bananas before preparing for drying. Discard any that are very ripe with obvious black spots or green on the peel. Peel and slice bananas thinly. Dip bananas in honey if desired before placing on dryer trays. Bananas may be sulfured for 45 minutes before drying. Process in dehydrator or oven, until bananas are pliable to crisp. Bananas do not sun dry well as they usually ferment before drying sufficiently. Dried bananas, once reconstituted, make wonderful banana bread or muffins.

Dried zucchini chips

Choose fresh, firm, young zucchini with shiny, smooth skin. Wash, and slice squash approximately ¼ -inch thick. Sprinkle slices with seasoned salt or dip in thin barbeque sauce. Spread slices on dryer trays and dry until slices are sufficiently dry and crisp. Zucchini chips that are dipped in a thin barbeque sauce will take longer to dry then will those sprinkled with seasoned salt. You may find that you need to place the barbeque dipped slices on plastic wrap for drying as well to make clean-up easier. When dry, pack in airtight bags and label. Use within 1 month unless stored in freezer.

Shredded zucchini

Shredded zucchini can be dried for later re-hydration and use in zucchini breads and muffins. Wash and sort firm, fresh, young zucchini with shiny, smooth skins. Shred finely and spread thinly on dryer trays. Try to avoid having clumps of squash anywhere on trays. Dry until squash is dry and no longer tacky to the touch. Dried shredded zucchini should be somewhat crisp rather than pliable. Store in sealed, labeled, airtight storage bags.

Tomatoes

Successfully dried tomatoes are prized for many Italian dishes and for making dried tomato powder. Occasionally dried tomatoes turn blackish, from oxidation and lack of sufficient acid. Tomatoes that turn black are not harmful to eat, but do not look very appetizing either. Some people solve the problem by storing dried tomato slices in olive oil. Such tomatoes, while generally delicious, also add extra fat to the diet and the fat called for in many recipes will have to be adjusted accordingly.

Choose firm, fresh, fully ripe tomatoes for drying. Tomatoes with a high acid content will yield a more satisfactory product then will sweeter varieties. Wash tomatoes, peel and core if desired, or leave them intact and slice them thinly. Arrange slices on dryer trays and process in dehydrator, oven or sun until nearly crisp. Dried tomatoes should feel tough, leathery, or crisp when sufficiently dried. Store tomatoes in airtight containers, packed as tightly as possible. You can dip each slice in olive oil, and allow the extra oil to drain from the slices before packing.

Chop dried tomatoes in your blender to make tomato powder to season soups and sauces. Powdered tomatoes reconstitute well to use in place of tomato sauce or paste.

Dried onions

Select firm, large, fresh onions for drying. Trim the ends of the onions and remove the papery shells. Slice into even slices no thicker then ¼-inch thick. Onions can be diced into ½-inch dice before drying if desired.

Process onions until they are papery dry and no longer sticky to the touch. Store in airtight bags inside another plastic, glass or metal container as onions have a tendency to pervade everything with their pungent aroma.

When processing, you may wish to place your dehydrator out of doors so as to avoid making your entire home smell like onions which will likely linger for days.

Dried onions are wonderful used as is in nearly any dish, and can be processed in a blender until powdered or flaked. Mix powdered onions with salt to taste for your own onion salt.

Re-hydrating dried produce

Many dehydrated fruits and vegetables can be consumed in their dried state. Most dried fruits and a few dried vegetables can be re-hydrated or reconstituted easily for use in various recipes. Dried vegetables re-hydrate more slowly than do dried fruits because they have lost more moisture content during the drying process.

Use only as much water as is necessary to cover the vegetables. Although boiling water shortens the re-hydration process, cold water can be successfully used. You should not put the dried vegetable directly into boiling water, however, as that will tend to make the vegetable tough when they are reconstituted. You should cover them in cool water, and then bring the water to a boil. Be sure to save any of the nutrient-rich liquid which remains from re-hydrating your fruits and vegetables for use in other foods, like soups, breads, and sauces.

Re-hydration can take anywhere from fifteen minutes two hours or more, depending on the food and the manner in which it was prepared for drying. If fruits or vegetables will stand for more than two hours during the re-hydration process, refrigerate them to prevent bacterial growth.

After re-hydration is complete, vegetables and fruits are ready to be used. Reconstituted dried apricots make delicious apricot bars, and reconstituted dried, diced potatoes make excellent home fries, and re-hydrated dried, shredded carrots make a wonderful carrot cake.

If you are using dehydrated carrots, corn, peas, or green beans in soup or stew, simply add enough cold water to the vegetable to cover and allow to soak for a few minutes. Then add the vegetables with their liquid to the soup or stew and allow to cook until vegetables are tender.

Fruit leather

Fruit leathers are old fashioned, healthy fruit snacks that are enjoyed for the chewy texture and delicious fruit flavors. It is made by pureeing fruit to a smooth, thick liquid which is then poured onto a flat surface, and subject to drying. As the fruit puree dries it develops a leathery appearance and texture. After drying is complete, fruit leathers can be pulled off the flat surface they were dried on and they retain their shape. They can be cut into various shapes with cookie cutters or they can be rolled up into a handy fruit roll snack.

When using fresh fruit in season, the drops or culls can be used very satisfactorily in making fruit leather. Simply cut away and bruised or discolored spots within the fruits and proceed according to the recipe directions.

You can use canned or frozen fruits to make fruit leather as well. For canned fruits, drain off the syrup in which the fruit has been canned and continue with recipe directions. For using frozen fruits, thaw fruit completely, then follow recipe directions.

After making fruit juices or jellies you can use the leftover fruit pulp to make fruit leather. Press the leftover pulp through a food mill to remove and the peels or seeds. Add enough fruit juice to flavor the pulp and bring it to pouring consistency. Continue with recipe directions for drying.

Basic instructions

Line a drying tray or baking sheet with heavy plastic wrap. You may have to tape the plastic wrap down to the tray or sheet with masking tape. Do not use aluminum foil or waxed paper. Fruit leathers tend to stick to these materials and you risk eating them along with your fruit leather! You can also use baking sheets that have been lightly coated with non-stick cooking spray.

Wash and prepare fruits according to recipe directions. Puree fruit in a blender until it is smooth. When mixing fruits for leather, combine fruits after pureeing. Add sweetening, spices or flavorings after fruits are pureed or mixed if mixing is desired.

Fully ripe fruit usually needs no further sweetening added. Exceptionally tart fruits can be sweetened with the addition of 1 Tbsp corn syrup or honey per quart of fruit puree. More corn syrup or honey can be added if fruit is still too tart to suit your particular tastes. Granulated sugar should not be used in fruit leather because it has a tendency to crystallize during storage making the fruit leather brittle. Brown sugar doesn't tend to crystallize as readily as white sugar does, but it imparts a stronger flavor to the leather. Brown sugar is very good if used in apple or pear leather, but is less desirable with other types of fruit. Try drying a small batch of leather using brown sugar to see if you like the flavor it adds to your fruit.

Artificial sweeteners can also be used to satisfactorily sweeten too tart fruits for use in leather.

Flavorings and spices concentrate in fruit leathers when dried, so use them sparingly. No more than ¼ tsp of spice per quart of puree is needed. You want to just taste a hint of spice to complement your fruit, not overpower it. Flavorings and extracts can be added at the rate of ¼ or ½ tsp per quart of puree, depending on tastes. Lemon, orange, or grapefruit peel can be added at the rate of 1 tsp per quart of puree.

Pour the prepared puree onto the drying surface. Slightly tilt the tray to help spread the puree evenly across the tray or sheet. You can also spread the puree on the sheet using a rubber spatula. Allow a 1-inch border to allow for further spreading during the drying process. Add any desired garnishes to the puree while it is still moist.

Dry the fruit puree until it feels leathery but still pliable. There should be no sticky spots on any part of the fruit leather when drying is complete.

Remove the fruit leather from the tray or cookie sheet while it is still warm. If fruit leather is allowed to cool it will be harder to roll up or cut. Cut leather into desired shapes with cookie cutters, or simply roll up entire sheet of leather into one long roll. Then, cut roll into desired sized pieces.

On occasion, fruit leather that has been dried in the sun will be dried to the touch on the top but will still be moist on the bottom. If this happens, simply turn the leather over and allow it to continue to dry until the leather is no longer sticky.

If your puree is too thin because your fruits are very juicy, or if you added too much juice while blending, you can slowly cook the puree over low heat to evaporate some of the liquid before spreading the puree on the trays for drying.

If your puree is too thick, simply add a bit of fruit juice or water to thin the puree to pouring consistency.

Apple leather

Sort, wash and core apples. Puree apples, with or without skins, in a blender with a small amount of water or apple juice added. Apples may also be cooked and then pureed if desired. Add ¼ tsp cinnamon to each quart of pureed apples if desired. A small amount of brown sugar will help make tart apples sweeter as well as to impart a more robust flavor to your leather. Add at the rate of ¼ cup per quart of puree. Increase brown sugar if very tart apples are used. Spread apple puree onto prepared dryer trays or cookie sheets. You can sprinkle finely chopped peanuts on top of the pureed apples if desired. Process in dehydrator, oven, or in sun until no longer sticky but still pliable. Package in airtight bags or vacuum containers, and label.

Banana leather

Wash and peel bananas. Puree in blender until smooth. Add ½ tsp lemon or lime juice, and ¼ cup white corn syrup to each quart of pureed bananas and mix well. Spread puree onto prepared dryer trays or cookie sheets. Sprinkle top of banana puree with shredded coconut if desired. Process in dehydrator, oven, or in sun until no longer sticky, but still pliable. Package in airtight bags or vacuum containers, and label.

Strawberry leather

Wash berries and puree in blender until smooth. Combine equal parts of strawberry pulp and apple pulp to improve texture of finished leather. Add ¼ cup white corn syrup to each quart of puree if strawberries or apples are very tart. 1 tsp lemon juice to each quart of puree will help retain some color in making strawberry leather. Spread puree onto prepared dryer trays or cookie sheets. Process in dehydrator, oven, or sun until leather is no longer sticky but still pliable. Store in airtight freezer bags or in vacuum sealed containers, and label.

Dried fruit chips

If you left your fruit leather to dry too long and it become brittle, don't throw it out! Simply chip off the brittle pieces or chop a larger piece up in a blender for use as cereal topping, or you can mix them with nuts for a quick trail mix.

Vegetable powders and flakes

Dried vegetables that have been chopped into flakes or into a fine powder are wonderful for seasoning soups, making a quick vegetable stock, seasoning salads, or whipping into sour cream for a delicious vegetable dip.

Dry vegetables that are to be used to make vegetable powders and flakes should be dried longer so that they are extra crisp and brittle. Be sure blender jar is completely dry before chopping because any moisture present will cause your dried vegetables to clump instead of flaking or turning to powder.

Process dried vegetables in blender starting with ½ to 1 cup of vegetables at a time. It is harder to flake or powder larger amounts of vegetables evenly.

Place the dried vegetable pieces in blender and set the blender on the chop setting and turn on for 5 to 10 seconds. Turn off blender and scrape down sides of blender jar with rubber spatula. Turn on blender on chop setting for another 5 to 10 seconds. Repeat until vegetables are sufficiently flaked or powders.

Fresh vegetables can also be pureed before drying, as for vegetable leather. Vegetable leather can then be flaked or powdered in the blender in the same manner.

Flaked and powdered vegetables can be combined with other seasonings such as salt, black pepper, red pepper, basil, etc. Homemade onion or garlic salt cannot be equaled for flavor, and the salt actually helps decrease the likelihood of the onion powder caking during storage. The best way to judge the amount of salt and other spices to mix into your vegetable powder or flakes, is to start with a little and then taste the finished product, adding more as needed. Remember, you can always add more spice, but it is often difficult to add more vegetable, so start out with a small amount and add more spice or salt as your tastes require.

Powdered or flaked vegetables can be added directly to most recipes, but you may want to slightly increase the amount of liquid called for in the recipe. A good general rule is ¼ cup liquid for each tablespoon of powder or each 1 ½ tablespoons of flakes. Some vegetables get stronger in flavor as they dry. Tomatoes are very concentrated in flavor when dry and 1 tablespoon of tomato powder equals 1 medium fresh tomato. Onions and garlic also both dry very concentrated in flavor, so use them sparingly.

For most vegetables, 1 tablespoon of

powder, or 1 ½ tablespoons of flakes equals 4 tablespoons of chopped fresh vegetable.

From a nutrition standpoint it is better to store your dried vegetables in the sliced or whole form they were dried in and chop into powder or flakes as you need it. Vegetable powders and flakes don't retain freshness and nutritive value as long as do whole or cut up dried vegetables. After you have chopped vegetables into powder or flakes try to use it up within a month. After a month quality deteriorates rapidly.

Recipes using dried fruits and vegetables

These recipes have been created especially for use with dried fruits and vegetables. Be careful in following the reconstituting instructions with each of these recipes in order to achieve the best product results.

Luscious apricot bars

2/3 cup dried apricots
½ cup butter
¼ cup sugar
1 1/3 cup sifted all purpose flour
1 cup brown sugar, firmly packed
2 eggs, well beaten

½ tsp baking powder
¼ tsp. Salt
½ tsp vanilla
½ cup chopped almonds

Rinse apricots; cover with water and simmer 10 minutes. Drain, cool and chop.

Combine the butter, white sugar and 1 cup of the flour; mix until crumbly. Pack into greased 9" square baking pan. Bake in 375 degree F oven for 20 minutes.

In a medium sized mixing bowl, gradually beat brown sugar into the eggs. Sift together remaining flour, baking powder and salt. Add the dry ingredients to egg mixture and mix well. Add vanilla, ¼ cup of almonds and the chopped apricots. Spread this mixture over the baked layer. Sprinkle the top with the remaining almonds. Bake in 350 degree F oven for 20 minutes. Cool in the pan on a wire rack. When completely cool, cut into 1½ inch squares. Sprinkle lightly with confection sugar, if desired. Makes 2 ½ dozen bars.

Delicious carrot cake

Boiling water
1 ½ to 3 cups dried shredded carrots
2 cups all purpose flour
1 cup granulated sugar
1 cup brown sugar, firmly packed
2 tsp baking powder
1 tsp salt
2 tsp baking soda
2 tsp cinnamon
½ tsp nutmeg
4 eggs
1 ½ cups vegetable oil
1 cup chopped walnuts or pecans
Cream cheese frosting (recipe follows)

Pour boiling water over dried carrots. Let stand to reconstitute about 25 minutes. Drain and measure 3 cups reconstituted carrots. Preheat oven to 350 degrees F (175 C). Generously grease and flour two 9-inch round cake pans or 1 large 9 x 12 x 2" baking pan.

In a large mixing bowl, combine flour, granulated sugar, brown sugar, baking powder, salt, baking soda, cinnamon and nutmeg. In a smaller bowl, beat eggs and oil together. Add to flour mixture. Stir 300 strokes or mix with an electric mixer on low speed until blended.

Fold in carrots and chopped nuts. Pour into prepared baking pans. Bake 25 to 30 minutes or until wooden pick inserted in center of cake comes out clean. Cool in pans on rack 10 minutes. Turn out onto rack to cool completely. When cool, frost with cream cheese frosting.

Cream cheese frosting

1 8-oz pkg. cream cheese
2 cup butter, softened
2 tsp vanilla extract
2–3 cups powdered sugar
1 cup chopped walnuts or pecans, optional

In a medium size bowl, beat cream cheese and butter until fluffy. Blend in vanilla. Gradually add powdered sugar. Beat until mixture is smooth and creamy, adding more powdered sugar as needed. Fold in chopped nuts, if desired, reserving a few to garnish the top of the cake. Spread frosting over top and sides of cake. Sprinkle nuts on top of the finished cake, if desired.

Home fried potatoes

2 cups dices dried potatoes
2 Tbsp dried onion flakes
Boiling water
½ tsp salt
¼ tsp black pepper
¼ cup vegetable oil or bacon drippings

In a medium saucepan, combine dried potatoes and onion flakes. Cover with water and bring to a boil. Remove from heat and let set 20 minutes or until tender. Drain. Stir in salt and pepper. In a large skillet, heat oil or drippings. Spread potatoes in skillet. Fry slowly, stirring occasionally, until potatoes are a golden brown. Serve piping hot with ketchup. Makes 4 servings.

Homestead vegetable beef soup

Meaty beef bone
2 beef bouillon cubes
5 cups boiling water
2 cups mixed dried vegetables such as green beans, corn, peas or potatoes
1 1-lb canned tomatoes, untrained
¾ cups pearl barley
½ tsp salt
¼ tsp pepper
½ tsp basil

½ tsp thyme
1 Tbsp dried parsley
3 Tbsp butter or margarine
½ cup chopped celery
½ cup chopped onion

Dissolve bouillon cubes in boiling water. Pour over beef bone in large soup pot. Bring to boil and reduce heat, simmering until meat can be easily removed from the bone. Return meat to water, and discard bone. Add vegetables and allow to simmer for an hour or until vegetables are nearly tender. Add tomatoes, barley, and seasonings. Simmer uncovered for 30 minutes.

In a small skillet, melt butter, and add celery and onion. Cook until celery is tender and onion is transparent. Add to soup. Simmer 30 minutes longer or until all vegetables are fork tender.

Seasoned salt

1 tsp tomato powder
1 tsp green pepper powder
1 tsp onion powder
1 ½ tsp garlic powder
2 tsp paprika
1 tsp dry mustard
1 tsp oregano
¾ tsp thyme

1 tsp celery salt
¾ cup pickling salt

Blend all ingredients except pickling salt in blender until fine. Add pickling salt. Blend only 1 or 2 seconds to mix. If mixture is too fine it will cake during storage. Store in a shaker that has an airtight lid. Makes 1 cup.

CHAPTER ELEVEN

MEAT SMOKING AND CURING

IN DAYS GONE BY, our ancestors preserved their meats by curing, pickling, and smoking. Today, we have the luxury of pulling fresh tasting meats directly from our freezers and have little need for the heavily salted pork of yesteryear. However, once you have tasted that first amber, smoked ham fresh from the smokehouse, you will start to be a believer in the ways of the pioneers. One taste of that delicately smoked, salt and sugar cured pork will convince you that you will never again be satisfied with the bland, flavorless hams or bacon from the local grocery store. You'll never want to be without your own backyard smokehouse again.

Home pickled corned beef is a treat that can be had on a regular basis, instead of waiting for a spring holiday to enjoy, and for a fraction of the cost as well.

The Smokehouse

There is a difference in cold smoking, and hot smoking. Hot smoking cooks the food as it smokes. The food smoked in a hot smoker can be consumed right away. Cold smoking is done through cool smoke, smoke that is under 120 degrees F. The food doesn't cook while it smokes, so it must be cooked before being consumed. Hams and bacons are put through the cold smoking process, and must be cooked or frozen after being smoked. Country style hams, more heavily salted than regular hams, can be left hanging in the smokehouse throughout the winter, but must be soaked overnight and cooked thoroughly before being consumed.

The smokehouse is a necessity if you want to create your own luscious hams and bacons on your homestead. The smokehouse can be a permanent affair constructed out of concrete block, wood or stone. An old shed can be converted to a smokehouse, or you can build one from scratch. An old wood stove can be set outdoors and the smoke piped into the smokehouse or a burner plate can be put into a corner of the room and a slow fire lit on the plate. Be sure to construct ample drafts in the top of the building to create proper air flow. The room should be allowed to get heavy with smoke, and the temperature should be maintained at between 80 degrees F and 120 degrees F. Higher than 120 degrees F and the meat will begin to cook. Lower than 80 degrees F and the temperature will be too low to ward off bacterial growth of the meat.

A less permanent but equally efficient smokehouse can built from such easily obtainable materials as an old, non-operating refrigerator or freezer, a metal trash can, and some pipe. Strip out the motor and compressor from the refrigerator or freezer. Strip out any plastic parts that could melt or interfere with the smoking process. Remove the lock assembly from the door of the unit to prevent the possibility of a child getting accidentally locked into the smokehouse. A hook and eye latch can be installed to ensure proper closing of the door. Cut a hole in the bottom of the unit, big enough to accommodate the diameter pipe you will be using to conduct the smoke from the fire to the unit. Cut a second hole in the top of the unit and insert a stove pipe cap and a butterfly damper. This damper will be used to regulate the airflow through the unit. Screw several eye-hooks into ceiling, inside of the unit to accommodate the meats. Put a thermometer inside the unit.

Approximately ten feet from the smokehouse, bury a short metal trash can or metal garden tub for use as a fire pit. You could even cut down a metal 55 gallon drum and use the bottom third of it as a fire pit. You can cut a sheet of tin slightly larger than the opening of the fire pit. Cut a hole in the top of the tin to accommodate the pipe. You can use 10 feet of stove pipe or clay water pipe to connect the fire pit with the refrigerator smokehouse. You'll need an elbow on either end of the pipe.

Conduct a dry run of your new smokehouse to be sure everything works properly and to familiarize yourself with the smoke conditions of your smokehouse. Lay a small fire in the fire pit using hickory, apple wood, or other hardwoods.

Never use pine or other soft woods to smoke meat as softer woods burn fast and hot, giving more heat and less smoke, and they can give your finished product and off flavor. You want a slow, green, smoky fire. Don't use paper or lighter fluid in your fire pit. Dry leaves and a handful of dry twigs or one or two pieces of hardwood kindling are all acceptable fire starters.

Light the fire in the fire pit, and when it is burning low, put the tin covering on the pit. The smoke will go through the pipe and fill the refrigerator smokehouse. Keep the butterfly damper and the door to the smokehouse both closed. Every hour or so, check on the temperature of your smokehouse interior. If it gets too hot, open the butterfly damper at the top, or toss a handful of wet leaves onto the fire to smolder it somewhat. If the smokehouse is running too cold, under 80 degrees F, you may want to stoke the fire a little bit. Adjusting the damper and fire will eventually give you the feel to create the perfect smokehouse environment for your cured pork cuts.

Before you can smoke your hams and bacon, you will need to cure or brine them.

Curing and Smoking Hams and Bacon.

There are numerous ways to cure and smoke hams and bacon. Salt may be used alone, with sugar, or with sugar and nitrite. The last method, sometimes referred to as "sugar cure," uses dry ingredients, or liquid ingredients, and sometimes, combinations of both.

You can also purchase premixed dry rub and pickle brine if you don't want to make your own or can't find the ingredients to make your own mix. The biggest disadvantage to using the premixed rubs is that it is harder to adjust them to individual tastes for saltiness or sweetness.

The dry sugar cure is safest if you

have no refrigerated curing room or equipment for brine curing. Mix up the curing ingredients as follows:

8 lbs salt, 3 lbs brown or white sugar, 2 oz saltpeter.

Use 1 oz of cure per 1 lb of pork for heavy hams weighing more than 20 lbs, use 1–1/2 oz cure per 1 lb of ham. Hams should be rubbed three separate times at three to five day intervals. Bacon should have one thorough rubbing with a light sprinkling over the flesh side after rubbing. Picnics and butts should have two rubbings at three to five day intervals. Rub the dry cure in well, getting the salt in all the folds and pockets of the meat.

Place the rubbed meats in boxes, on shelves, on wooden tables to cure but not in tight boxes or barrels where they rest in their own brine. Do not use cardboard or galvanized containers. The length of curing should approximate seven days per inch of thickness. For example, if the ham weighs approximately fifteen pounds and is approximately five inches thick through the thickest part, this ham should be cured thirty five days. If bacon is two inches thick, it should be cured for fourteen days. It is important to be sure to rub some of the curing salt into the aitch bone joint and hock end of ham to guard against a condition known as bone sour.

It won't hurt to leave the product in cure longer than the recommended time since the saltiness does not increase. Do not under-cure.

Dry curing should be done in a cool place to reduce the risk of spoilage. If you have a secondary refrigerator, you can keep the curing meat in the refrigerator bins for the duration of the curing process. Check the curing meats every few days and drain off any liquid that collects in the bottom of the refrigerator bins. After curing, wash the hams and bacon with a stiff brush under cold running water to clean off excess salt before smoking.

Pickle brining involves curing meat in a liquid cure solution for several days before smoking. The salt component in brine is the primary one for preservation. Sugar is added to counteract the harsh flavor of the salt, and spices to taste. Saltpeter will help produce a rosy pink product, but can be avoided altogether if you prefer. Non-iodized salt without any chemical additives is a must. Brown sugar adds the most flavor, although white sugar is used most often. Molasses and honey can also be used to sweeten the pickle. Some good additions to the pickle might be a spice bag or tea-ball containing bay leaves, cinnamon, nutmeg, cloves, juniper berries and thyme. Other excellent additions are wine, cider, and brandy. Keep in mind that it is important to keep the salt concentration high enough to preserve the meat and retard any spoilage, so if you add any of these liquids, you must add additional dissolved salt to compensate for the increased liquid. Mix up the pickle brine with these ingredients:

6 lbs salt, 2 lbs brown or white sugar, 2 oz saltpeter, 4 1/2 gallons of water.

Bring all ingredients to a boil and simmer until all ingredients are dissolved and mixed. For a milder pickle cure, increase sugar by one pound and add one more gallon of water.

Wash a 5 gallon plastic food grade bucket, wooden or plastic barrel, or stoneware crock with hot soapy water. Rinse with boiling water, and dry the bucket well. Never use a metal bucket as the brine can react with the metal causing corrosion. Pour the hot brine mixture into the bucket. Allow it to cool. Put your meat in the brine, making sure the entire surface of the meat is beneath the brine. Place the brine buckets or barrels in a cool place with a temperature between 35F and 40F. Large cuts of meat can be left in brine for up to three or four weeks, smaller cuts up to a week, before becoming too salty. .

After 5 or 6 days, it is wise to remove

any of the smaller pieces unless you want them to become very salty. The larger pieces will be nicely cured in one week. You can leave meat in brine for as long as 5 weeks, if you continue to replenish the solution, but it will be very salty.

Before placing any un-boned joints in brine, you should prick them with a long needle down the length of the bone, several times, to help the cure permeate. You can purchase tools called brining pumps which will allow you to inject brine curing down to the bone at some butcher supply stores. You can also purchase marinade injectors at most houseware departments. These will work fine for homestead use. Keep the pump or injector sterile, and be careful not to inject air into the meat, which can cause pockets of spoilage.

When the meat is thoroughly brined, wash it in cold running water to remove excess salt. Hang it up to dry thoroughly if you are planning to smoke it.

Combination cures are rubbed on and weighted down to draw out the natural brine within the meat. A standard combination cure recipe is: 6 lbs salt, 3 lbs sugar, 2 oz saltpeter (optional). Spices such as cloves, nutmeg and black pepper can be added to taste. Rub the meats thoroughly with the dry cure and bury the meat in dry cure in a water tight wooden barrel or plastic bucket.

Press the meat beneath a weight. The pressed meat will eventually exude a liquid, which will turn the dry cure to a very intense brine. You can stretch this brine a bit by adding flavorful liquids like cider, brandy, wine or maple syrup. Leave in this brine for 5 days for bacon and smaller cuts, or up to 4 weeks for larger cuts.

A brine cure will usually produce a milder result than a dry cure, though you can improve a dry cure by allowing drainage of the strong brine formed by the escaping meat juices and replacing it with savory liquids such as cider, wine, brandy or maple syrup. The middle of the road alternative is to do no drainage and to add additional fluids or not, depending on your taste.

A general guide for brining time is: Bacon—1 1/2 days per pound of bacon. Hams and shoulders—2 days per pound. Longer cures would be appropriate with more primitive methods of storage, or in making what is known as country ham, which requires no refrigeration after smoking.

After brining is complete, wash under running water to remove excess salt and brine, and hang up to dry before smoking.

Smoking

After the cuts are dried, brush them with pineapple juice or maple syrup. Let dry slightly. Hang meats in smokehouse after a good, smoky hardwood fire is smoldering in the fire pit. Smoke bacon for 24 to 36 hours, depending upon your tastes. This can be accomplished by smoking for a straight 24 to 36 hours, or for 2 or 3 12 hour days, or by smoking for 6 hours a day for 4 to 6 days. If temperatures exceed 40F when not smoking, it is best to bring in the meat and refrigerate it until you are able to continue smoking.

When the meat is a rich amber to chestnut color, it can be taken out of the smokehouse, sprinkled with a little red pepper, and wrapped for the freezer. Bacon can be sliced before freezing or can be frozen in slab form, as is desired.

Smoke Hams for approximately 36 to 48 hours, depending upon your tastes. This can be accomplished all at once, or by breaking up the smoking time over a successive period of days. Again, if the temperatures exceed 40F when not smoking, it is best to bring in the meat and refrigerate it until you are able to continue smoking. When the hams are a rich amber to chestnut color, they can be taken out of the smokehouse. Sprinkle hams with a bit of red pepper and wrap for the freezer.

Cured and smoked meats should not be kept frozen for extended periods of time. After a few months the quality and flavor deteriorates.

Corned beef

Boneless brisket, plate, chuck, or beef are usually used for corning. For each 10 lbs of meat dissolve in ½ gallon of hot water, 1 lb of salt, 1/2 lb of sugar, 1/2 oz of baking soda, and 1/3 oz saltpeter or ½ oz cream of tartar. (5–1/2 tbsp of a curing salt containing 0.5 percent sodium nitrate and 0.5 percent sodium nitrite may be used as replacement for the salt, and saltpeter or cream of tartar). Garlic and pickling spices (cloves, peppercorns, bay leaves, and thyme) may be added in varying amounts if more flavor is desired. Chill brine to cool.

Place the meat in a stone crock, plastic bucket, or wooden tub (do not use metal containers that will corrode). Put the chilled curing ingredients in the container in sufficient quantities to cover the meat. If using garlic and pickling spices, add them, stir, and weight the meat with a board upon which a nonmetal weight can be placed. If the cut meats are not more than 3 inches thick in the thickest part, they will cure in approximately 12 to 14 days. Remove from brine, and wrap for the freezer.

CHAPTER TWELVE
HOUSEHOLD CLEANING HINTS

TODAY'S HOMESTEADER HAS THE need to be thrifty with, both money and time, as well as the desire to be health conscious. Making your own cleaning supplies is a good way to lower your household expenses, and help to ensure your family's health by doing away with commercially produced chemical cleaners for your home.

You may wonder if homemade household cleaning agents work as well as their commercial counterparts. The answer is a resounding yes! They work as well, or even better in many cases, and the homemade agents don't have the health hazards associated with them that the commercial cleaners often have. Most homemade cleaning agents are composed of things found right in the average homestead every day.

Wood Furniture

Mayonnaise can be used to restore and condition wood furniture. It is especially good for removing white water rings on furniture. Spread a thin coating of mayonnaise on your wood furniture and let it set for about 10 minutes. Gently polish off the mayonnaise with a soft cloth. For particularly stubborn water rings you may have to leave the mayonnaise on the spot overnight and polish off in the morning. Mayonnaise will also remove crayon marks from wood furniture.

Glass cleaner

Take a spray bottle and half fill it with ammonia and vinegar, using ¼ cup vinegar per ½ cup ammonia. Fill bottle the rest of the way with water. Use this to wash windows and mirrors, wiping with newspaper for a streak-free shine.

For a heavy duty window cleaner to use on those really dirty windows, mix ½ cup ammonia with ½ cup white vinegar in a bucket. Fill bucket with warm water and add 2 Tbsp cornstarch. Mix well.

Tile, Vinyl or laminate floors

Mix ½ cup vinegar with ½ cup ammonia in a pail or basin. Add 1 Tbsp dish detergent, and 1 gallon of warm water. This is a great floor cleaner that cuts through grease and grime, and dries virtually streak free .

Remember this; ¼ cup oxygen bleach mixed with a gallon of hot water will help restore any dingy floor to its original brilliance. Oxygen bleach is also good for keeping the grout between ceramic tiles clean.

Most black heel marks can be instantly removed by rubbing them with an ordinary pencil eraser.

Carpets and upholstery

Shaving cream makes a very effective carpet and upholstery spot remover. Plain club soda is also a great stain remover for carpets, especially for new spots. Rub a little on the spot, let it set for a few minutes, and then sponge it up completely. For the best treatment, rub stains with shaving cream, blot up cream with sponge or cloth. Then rinse spot with club soda, and blot thoroughly.

Sprinkle salt on muddy footprints in your carpet. Allow about 15 minutes for the salt to work, then simply vacuum up the dirt.

Countertops

Hot soapy water is the safest agent to use to clean your kitchen counters. You wash your dishes with soap and hot water, why not your countertops? Hot soapy water has anti-fungal and anti-bacterial properties, perfect for use in kitchen and bath.

The real key to clean work surfaces in the kitchen is to use a clean cloth for each cleaning. Don't use a sponge or cloth that was previously used and dried. There will be bacteria and fungus present in a cloth that was previously used, and by using it again you will only be spreading that bacteria or fungus around your kitchen further.

Too many commercial kitchen cleaning products have poisonous chemicals that we wouldn't dream of eating, yet we routinely use them to clean our food preparation surfaces. Any chemical to food transfer, no matter how minute, is undesirable and potentially dangerous.

Appliances

Clean your refrigerator, freezer and washer with a clean cloth soaked in a strong solution of bicarbonate of soda and hot water. You can then wipe out the refrigerator with another cloth dipped in a solution of water and vanilla essence.

Running your empty dishwasher through a rinse cycle with ½ cup of vinegar added will help get rid of hard water deposits, odors, and food stains. You can also run your empty dishwasher through a wash cycle after filling the detergent dispenser with lemonade mix. The lemonade cuts any hard water deposits, and leaves your dishwasher smelling clean and fresh.

Bath fixtures

A strong soap and water solution is good for cleaning bathroom countertops and basins.

A tablespoon of commercial lemonade mix mixed with hot tap water in a bowl can be used to clean a plugged up showerhead. Just let the showerhead soak overnight, and in the morning rinse it clean and replace it in your shower. The lemonade easily dissolves the hard water deposits that can clog your showerhead.

To clean soap-scum ring off the bathtub use a scrub brush and a strong dish soap and hot water solution. Dip the brush in hot water and lightly scrub along the ring. Then, rinse tub with hot water. A car washing mitt and a liberal sprinkling of automatic dishwashing detergent will cut through the soap scum to make the tub or sink shiny and clean.

Put 1/8 cup of oxygen bleach powder in your toilet, brush with a toilet brush, and flush. If this is done on a weekly basis, it will keep your toilet shiny and white. Lemonade mix, used the same way will also keep the bowl sweet smelling and sparkling clean. Automatic dishwashing detergent can also be used in the same manner to clean toilets.

Miscellaneous cleaners

You can use eucalyptus oil to remove the sticky residue left from stickers and labels on non-porous surfaces throughout your home. Just use caution when using eucalyptus oil as it is very strong and can cause irritation. Be sure to wash your hands thoroughly after using eucalyptus oil.

Rust stains can be removed in several ways. You can apply a paste made of salt and lemon juice, borax and lemon juice, or cream of tartar and water.

Apply this paste to the rust stain and let it set for 30 minutes. Then wash off and rinse with clean water. Repeat application if necessary. Hydrogen peroxide also removes rust stains. Pour it on the stain, let is set for 15 minutes, and wash and rinse as usual. Repeat application if needed. A paste made from an oxygen bleaching agent and water can also be applied to the stain, then washed off after 15 minutes. Rinse tub well with hot water afterwards.

Indoor and Outdoor pests

Boric acid, sprinkled liberally in any crevices and cracks will kill cockroaches. It doesn't work as fast as some commercial pesticides, but it is safer, and works for a much longer time period, up to a year usually. Boric acid also works on silverfish, especially if mixed with a small amount of sugar, and sprinkled around affected areas.

Toss a couple of bay leaves into your grains and flour to help prevent insect infestations. Laying a few sprigs of spearmint leaves or a few sticks of spearmint chewing gum around your pantry or storage shelves will also help deter mealworms.

Should you have a problem with raccoons and stray cats or dogs attacking your garbage cans, sprinkle the inside of each can, including the lid, with ammonia or black pepper. Most animals hate the smell of these products and will avoid those cans.

Moles can be discouraged from tunneling into your gardens by putting an old rag soaked in olive oil into the mole hole. Likewise, spearmint chewing gum stuck in the holes will also deter these pests. They will usually turn around and tunnel in a different direction rather than pass the oil-soaked rag or the mint.

Beer is a wonderful pest control agent. Bury several empty tuna cans up the rim in your garden soil. Fill them with beer and leave them overnight. Check each trap and empty them every morning. Various slugs and insects can be trapped and disposed of in this manner.

Try never to use poisonous mouse and rat bait around your homestead. Mice and rats may eat it, but they either go off to die, usually within the home's walls, creating a nasty odor that persists for weeks, or they go outdoors to die and are in turn consumed by free-ranging chickens or the family cat or dog. Needless to say, that is a good way to kill poor Fluffy, or even a couple of homestead chickens.

The best and safest way to take care of a rat or mouse problem is with a two-fold plan of attack. The first part is to cleanup anything that might be attracting the rodents into the area. Feedbags carry the scent of grain and are big rodent attractors. Keep feed in clean metal trash cans with tight fitting lids. Secondly, use either mechanical traps or sticky traps to catch and kill rodents. Mouse traps should be set in the garage, basement, and food storage rooms, as well as anywhere else in the home or barn where mice or rats have been spotted. Be sure to check traps often and dispose of any rodents properly.

Seal up any holes where rodents might gain entrance to your home or barn. Rodents are not only a nuisance, but carry several serious diseases that can affect humans. Get a good cat. A good cat is a valuable asset to any homestead. Cats not only catch mice, but their very presence often discourages new mice from entering their living space. Because of the open nature of most barns, a barn cat or two can be vital in the control of rodents in the barnyard.

CHAPTER THIRTEEN
PRINCIPLES OF FOOD STORAGE

EVERY HOMESTEAD NEEDS A plan for food storage. Where are you going to put all those bottles of fruits and vegetables that you so carefully preserve? Do you have space for the freezer in your laundry room or kitchen? How about all those root crops and winter squash? Those five gallon buckets of honey will demand a bit of space as well. You will also need a cool, dry place to age and store your homemade cheese.

This is where the store room comes in. In the olden days, it was called a root cellar, but today that room does so much more than merely store root crops. You can successfully create a multi-purpose food storage room with a little thought and pre-planning.

You may find you need to build on a small room for storage. The value to your family will make the investment a small one by comparison. A dry corner of your basement that can be walled off from the rest of the room is ideal. Don't try to use a corner of your utility room because any heat source that is present will only upset the temperature/humidity ratio and aid in the rapid deterioration or outright spoilage of your food stores. A hot water heater puts out a surprising amount of heat, and a clothes dryer puts out humidity as well as heat. Even a freezer puts out heat into the environment, so you should not put your freezer in your store room. It is best to incorporate a freezer into your utility or mud room, or even into your kitchen or garage.

A closet that is on an outside wall can be useful for limited food storage. Canned goods and dry goods can be successfully stored long term in a closet, however, root crops, squashes and fresh fruits will quickly deteriorate in such a warm environment. If using a closet, these fresh foods can be kept in the refrigerator until used.

A secondary refrigerator is a real boon to the homesteader. Used refrigerators can be picked up at very little expense and are well worth the cost. Put a secondary refrigerator in your basement, garage, or shed, and use it to age and store cheeses, extra milk, and eggs. This refrigerator will also be the perfect environment to cure hams and bacons before smoking them.

Be sure to have ample space for water storage, whether it be a water barrel, or jugs, or cases of water. Fresh water storage is one of the most often overlooked facets of food storage on the homestead. Water can also be put into clean, empty soda bottles and placed in your freezer. The water bottles can be thawed out and used for fresh drinking water if supplies run short, and the frozen water also helps to keep the freezer colder during any power outages. Remember, a full freezer will keep food frozen longer during power outages than will a partially full freezer. You also may want to consider investing in a small generator to maintain your freezers during a power outage.

You'll want shelves and lots of them. Wooden shelves are sturdier then the wire ones sold in hardware stores and home improvement centers. Sometimes you can purchase commercial shelving from a business that is closing or moving. These commercial shelving units are ideal if the price is right. Some of your shelves can be used for storing clean, empty canning jars and supplies. Other shelves can be used for storing full jars, and cans, as well as boxed and bagged dry goods.

A pallet placed on the floor is adequate for short term storage of 25 or 50 pound bags of flour, rice, beans, and sugars. You must be diligent in deterring rodents and insects if you chose to store your bagged goods for any length of time in this way. A few sticky mouse and insect traps scattered here and there in the store room should work to keep most pests out of your grains and beans.

Bins with wire mesh bottoms are perfect for storing root crops. If you prefer, you can store your root crops in hanging mesh bags. A planter type box for sanding carrots and other root crops is nice to have as well.

You will also want a large, removable shelf for aging homemade cheese. One that can be taken down and scrubbed clean, then sun-dried to inhibit mold growth is preferred. Another shelf for curing soaps will be a plus to the homesteader who plans on producing home-made soaps.

If possible, you should procure a few used kitchen style cabinets to put in the food storage area for dust-free storage of food processing supplies, such as caning supplies, grinders, and canners, as well as soap-making and cheese-making supplies.

Hooks or nails stuck in the walls are

great for hanging onion or garlic braids. Hooks screwed directly into the ceiling are also important for curing and storing dried herbs.

As you can see, your homestead's food storage room is a vital part of your food storage plan, and your food storage will be an extremely important part of your self-sustaining lifestyle.

Produce bins

Potatoes, rutabagas, sweet potatoes, apples, pears, and onions can all be stored in wooden or mesh bins. Never store apples in close proximity to potatoes or other root crops, as apples give off a gas that hastens the ripening process of other vegetables, leading to faster spoilage of stored crops. It's better to store your apples in a spare corner of your basement then to risk ruining your other stored crops.

These bins can be built in any one of several different ways. A good size for one bin is 2x2x3 feet. Make two square frames using 1x4 lumber. Staple heavy hardware cloth to the underside of one frame. Cut slats that are approximately 1 ½ inches wide by 3 feet long. Space them about ¼ inch apart, vertically on the frames, spacing the frames an equal distance, about 2 feet, apart. Be sure to set the frame with the hardware cloth attached to the bottom of the bin. Each bin will stand on the ends of the slats, the bottom of the actually bin space will be kept up off the ground by about 6 inches of slats.

You can also build your bins nearly completely out of heavy gauge hardware cloth. Cut and nail your frames the same as above. Staple hardware cloth onto the bottom of one square frame. Attach 4 vertical braces to the corners of your frames, connecting the frames together. Stretch heavy gauge hardware cloth over this framework, stapling firmly in place with heavy duty staples. These bins will stand on the vertical legs produced by the frame, keeping the mesh bottom off the ground. These bins are not quite as sturdy as the wooden bins, but they are a bit more impervious to mice.

Sand box for carrots, beets and other root crops

Fresh carrots can be had year round by harvesting and re-planting your carrots and beets in a box full of clean, moist sand in your store room. You can build a rectangular wooden box very easily, or you can purchase window boxes in which to sand carrots. Trim the tops back before planting the roots in the moistened sand. The tops will continue to grow back and the roots will stay crisp and delicious.

CHAPTER FOURTEEN
Natural Remedies

Nature provides us not only with wild, natural foods free for the picking, but also with natural medicines, or remedies, also free and in abundance. Natural remedies have a place in treating many illnesses, as first aid treatment, as well as in conjunction with other conventional medical treatments. Many natural remedies can be found in the wild, several of them can be grown in the home herb garden, and a few may be purchased in your neighborhood health food store or co-op. A great many of our modern drugs are derived from herbs, as well as the bark and roots of various trees and plants.

Remember that natural remedies are often effective within hours, and many natural alternatives to commercially produced drugs are equally effective, with fewer side effects, and are generally safer, however, they can be just as potentially dangerous if abused or used carelessly. Also, if your condition fails to improve after twenty four hours of treatment with natural remedies, or worsens or reoccurs after treatment with natural remedies it may be necessary for you to visit your doctor for more aggressive treatment.

Even when used properly, some natural remedies may produce allergic reactions in sensitive individuals. If this occurs, discontinue use and consult your physician for further treatment.

There are a few warning symptoms that should only be evaluated and treated by a qualified physician. If you experience any of these symptoms, please do not attempt to treat them with any natural remedies as they may be symptoms of a more serious condition. These symptoms are: Severe head, chest or abdominal pain. Severe head pain may be a symptom of meningitis, or aneurysm. Pain in the chest could indicate heart problems, and pain in the abdomen could be a sign of appendicitis, tubal pregnancy, etc.

Both fresh and dried herbs can be used in producing natural remedies. Fresh herbs are more potent than dried herbs, however, dried herbs have the advantage of being available year round. Homegrown herbs are easy to dry and easy to store. Properly dried and packaged, dried herbs will keep potency for a little over six months. Tree bark, plant roots, and seeds will keep potency for several years if stored properly. Remember, when using fresh herbs, you will need approximately twice as much as you would the dried herb.

A small kitchen scale is very handy in the preparation of herbal remedies. A small mortar and pestle are also important tools to have on hand in your homestead kitchen.

Herbal remedies can be prepared for use in several different ways. There are infusions, decoctions, tinctures, poultices, and compresses.

Infusions

An infusion is produced by steeping herbs in water, as in making tea. Herbal infusions are often referred to as herb teas. A standard infusion is made by placing 1 ½ ounces of the desired dried herb, or 3 ½ ounces of the fresh herb, in a glass or china teapot. If using a combination of herbs, the combination should equal a total of 1 ½ ounces dried herbs or 3 ½ ounces of fresh herbs. Pour 24 ounces (3 cups) boiling water over herbs in pot. Let this steep for 15 minutes. Strain infusion through a very fine sieve or a small piece of muslin cloth.

Drink the infusion while still warm. The standard dosage of an herbal infusion is one 8 ounce cup 3 times a day. A bit of honey may be added to unpleasant tasting herbal infusions to make them more palatable. Some infusions can be added to bath water, or to a foot soak, or used as a hair rinse or face wash.

Bath infusion bag

Cut out a 8x8 inch square of muslin cloth. Place a handful of the desired herbs in the center of the square and gather the corners together. Tie with a ribbon and hang over your bathtub faucet so that hot water will flow through the bag as the tub fills. You can also let the bag soak freely in your bath water for a few minutes before you get into the tub.

Decoctions

Decoctions are used instead of infusions when woody plant parts, bark, or roots are used. Place 1 ounce of dried herb (root, bark, etc) or 2 ounces of fresh herb in 24 ounces (3 cups) of water, and bring to a boil. Allow mixture to simmer for at least 20 minutes, or up to one hour. Strain the decoction through a fine sieve or muslin cloth while still hot. If kept in a cool place in a tightly covered container, the decoction will last for 24 to 48 hours. The standard dosage is to drink 1 cup three times a day. Honey may be added to unpleasant tasting decoctions to make them more palatable. Some decoctions can be added to bath water, or to a foot soak, or used as a hair rinse or facial wash.

Tinctures

Herbal tinctures are mixtures containing the active compound of an herb dissolved in alcohol. It can be used topically or taken internally. A tincture can be taken as is, or can be diluted with a small amount of water. Tinctures have a longer shelf life then either infusions or decoctions, remaining potent for at least a year.

To make a tincture, take 2 cups of alcohol, at least 30 proof, (vodka works well) and pour it over 4 ounces of dried herbs (8 ounces fresh herbs) in a wide mouth bottle. Keep the mixture in a warm place for 2 weeks, shaking mixture daily. Strain the liquid through muslin into a dark, airtight bottle. The standard dosage is to take 1 tsp 3 times daily.

Compresses, poultices and plasters

Compresses and poultices are used to apply herbs externally. They are excellent for the treatment of headaches, muscle aches, coughs with congestion, fevers, and skin problems. It is easy to seal in the active warmth of a compress or poultice by laying plastic wrap over the cloth after it has been applied to the body. A hot water bottle or warmed gel-pack works equally well.

To make a poultice, boil a handful of the desired fresh herbs in a little water for 2 minutes or longer if the herbs are dried. Quickly spread the mixture on a strip of muslin cloth, and apply it to the body while the poultice still hot.

A compress is made the same way, except instead of using the whole herb, a cloth dipped in an infusion or a decoction is used.

Plasters are beeswax impregnated cloth strips that can be used to treat coughs and sore throat. Keep plasters in place with a strip of gauze. Heat 10 drops of desired essential oil, such as eucalyptus, thyme, or lemon, together with 1 fluid ounce of vegetable oil and ½ ounce of beeswax. Melt together over medium heat. Once the wax has melted, remove it from the heat and let the mixture cool until it has the consistency of a creamy paste. Spread this paste onto a strip of muslin cloth. Wrap the strip around the chest and secure with a long strip of gauze. The wax will continue to warm and soften with the body's heat, slowly releasing the essential oil.

Poultices and compresses should be as hot as is tolerable without causing burns when applied.

Ointments

Herbal ointments are also used topically. They work by forming a protective, healing layer on the skin and are used as beauty treatments as well as for healing skin or muscle problems.

The easiest way to make an ointment is to simmer 2 Tbsp of the desired dried herb in 7 ounces of melted petroleum jelly for about 10 minutes. Stir mixture up well then strain through a piece of fine gauze, being sure to wear rubber gloves to protect your hands from the heat. While still hot, pour the liquid ointment into a heat-proof container. Allow ointment to cool, and seal the container. Ointments keep potency for several weeks to a few months.

A few popular herbs

- Dill seeds can be chewed to help eliminate bad breath and as an aid to digestion.
- Black Walnut nut rinds, when used still green, make a very good treatment for ringworm.

- Spearmint, peppermint, catmint are all mints that help calm an upset stomach if taken in a soothing infusion or tea. Mint tinctures and infusions are also very good for treating the discomforts associated with gas.
- Chamomile is used in an infusion to promote relaxation, or to calm a queasy stomach.
- Ginger tea is another good remedy for upset stomach. Combining ginger, chamomile, and mint in a standard infusion is the best way to soothe nausea. Ginger and cinnamon can be steeped together in an infusion to promote good circulation. Drink 4 ounces (½ cup) of this tea 3 times daily. You can quadruple the recipe and mix it with 8 ounces of Epsom salts to create a delightful bath salt for an invigorating foot soak.
- Sweet flag, or Calamus root is a well known aid in the relief of stomach problems such as excess gas, and indigestion. Interestingly, smokers who chew the dried root of sweet flag usually find that it causes mild nausea, thereby making the plant a useful aid in breaking the smoking habit.
- White willow and Pussy willow bark contains salicin, an aspirin related compound that will help reduce fevers

and alleviate headaches.. Drink it in an infusion or decoction.

- Arnica Montana helps to reduce muscle aches and inflammation, as well as reduce the soreness of wounds when used in a compress or as an ointment. (Arnica may cause blistering in sensitive individuals)
- Tea tree oil, or Melaleuca has antibiotic, antifungal and antiviral properties. It should be used topically only as Tea tree oil poisoning has occurred in animals and humans. Tea tree oil can be used on insect bites and acne, as well as for the removal of skin tags, warts, and other such skin blemishes.

Other natural agents

- Honey has long been known for its antibiotic and antifungal properties. Honey is useful in soaps and lotions as well as in salves and ointments. Honey can also be used fresh, in its natural state, on wounds to promote healing.
- Oatmeal can be used to soothe sore, inflamed, or irritated skin, whether from poison ivy or insect bites. Just put 1 cup of oatmeal in a cheesecloth bag, attach it to your bathtub faucet and let the warm water flow through the bag. Or, you can grind 1 or 2 cups

of oatmeal to a fine powder in your blender, and add this powder directly to your bathwater. Oatmeal soap is good to keep on hand for use by those with sensitive skin.

Formulas

To treat diarrhea

1 tsp oak bark
1 tsp horse chestnut bark
½ cup cold water

Mix both barks and water in a small saucepan. Heat to boiling and boil for about 5 minutes. Take unsweetened, in mouthful doses until diarrhea subsides.

A natural laxative tea

3 tsp ground senna leaves
1 ½ tsp ground buckthorn bark
½ tsp ground psyllium seed husks
Pinch of powdered sassafras root bark
¼ tsp ground dark anise seed
¼ tsp ground blonde psyllium seed
¼ tsp ground dark fennel seed
¼ tsp ground allspice

Mix all ingredients very well. Add ½ tsp to 1 cup of boiling water, and allow to steep for 15 minutes. Strain, and enjoy the unusual flavor of this cleansing tea.

To treat influenza

1 tsp juniper berries
1 tsp coltsfoot
1 tsp lance-leaf plantain
1 tsp black elder flowers
1 tsp willow bark.

Mix all ingredients. Crush, and steep 1 tsp of this mixture in ½ cup of boiling hot water. Sweeten with honey, and drink 1 cup daily, broken up in mouthful-sized doses.

A good liniment

1 cup oil of camphor
1/3 cup oil of cloves
½ cup oil of wintergreen
½ cup oil of eucalyptus
½ cup oil of origanum

Mix all ingredients thoroughly, and pour into bottle or jar with lid. Use for soreness, swelling, pain and stiffness, colds and congestion, etc. Always shake bottle well before using.

Other natural remedies

For bee stings, apply a paste made from baking soda and water directly onto the sting. A slice off a small, fresh onion held in place by a length of tape is another well known cure for bee stings.

Mosquito and black fly bites are treated by rubbing wet bar soap directly onto the site to help relieve itching. A paste of baking soda mixed with lemon juice, vinegar or witch hazel is also useful for insect bites.

Splinters in fingers can be removed easier if the finger is first soaked in vegetable oil for a few minutes.

Treat a sunburn by lightly rubbing apple cider vinegar onto the affected skin.

Natural cosmetic preparations

These natural cosmetic compounds are far superior to their commercially available chemical counterparts. Try them, and see for yourself!

Cleansing almond mask

¼ cup almonds
Warm water

Grind the almonds finely in a blender or grinder. Mix in a small amount of warm water. Apply this paste to your face, and allow it to dry for 15 minutes. Wash off the dried mask with clean, warm water, followed by a cool water rinse. Gently pat your skin dry.

Apricot facial mask

Moisturizes and nourishes your skin

2 fresh or reconstituted dried apricots
Warm olive oil

Mash apricots well, and mix with enough oil to make a smooth, spread-able paste. Smooth this paste gently over skin and leave on to dry for 15 minutes. Wash off with clean, warm water, followed by a cool water rinse. Gently pat skin dry.

Moisture rich avocado mask

1 ripe avocado

Pit, and peel avocado. Mash avocado well, and warm it gently in a small saucepan over medium heat until just warm. Spread avocado over skin and let dry for 15 minutes. Wash off mask with clean, warm water followed by a cool water rinse. Gently pat skin dry.

Herbal beauty baths

A small muslin, draw string bag makes a good herbal bath bag. It can be turned inside out and rinsed out between baths, and can be safely tossed in the washer with your other laundry. Hang the herb filled bag over your water faucet so the warm bath water flows freely through it, or tie it tightly shut and float the bag right in your bath water.

Any of the following herbs can be mixed to create your own personal bath mixture.

- Calming herbs: fragrant valerian, lavender, lemon balm, chamomile, passion flower
- Skin softening herbs: rose petals, elder flowers, linden flowers, peony petals
- Healing and soothing herbs: mint, chamomile flowers, rosemary, elder flowers, linden flowers, yarrow flowers, lovage.
- Skin toning herbs: lavender, mint, thyme, yarrow flowers
- Invigorating herbs: cinnamon, mint

CHAPTER FIFTEEN
THE KITCHEN HERBS

MOST PEOPLE ARE MORE familiar with herbs grown for culinary use than for any other purpose.

The majority of homesteaders are growing common sage for use in fresh pork sausage, various mints for use in soothing teas, basil, chives, and parsley for use in the kitchen. Other herbs can be grown with equal success in the kitchen herb garden; herbs, such as savory, marjoram, thyme, tarragon, and oregano. And don't overlook such useful herbs as Rosemary, and horseradish. Some of these herbs have very specific growing requirements, many of which are addressed in the volume on gardening in this series.

All of the mints; Lemon balm, peppermint, spearmint, chocolate mint and apple or pineapple mint can be dried for use in teas. Dried peppermint or spearmint leaves can also be placed in a tea ball and steeped in a pot of hot chocolate to produce the delightful treat called minted hot chocolate. Fresh mint leaves can be frosted with sugar and used as a delightful, edible garnish for many desserts and beverages.

Many, if not all, of the garden herbs commonly grown on homesteads can be put to delicious use in herb jellies. Most of us are familiar with the sweet flavor of mint jelly, usually served with lamb or other meats. Basil Jelly is heavenly spread on a bagel with a little cream cheese, or slathered over garlic toast with any Italian dish. Marjoram Jelly is another great herb jelly that goes well with many meats including fish and pork. Don't be afraid to experiment to find new flavor combinations your family will love!

Frosted Mint Leaves

3 dozen large fresh mint leaves, any variety desired
1 egg white
Super-fine granulated sugar

Select perfect mint leaves with no holes or other insect damage. Wash leaves carefully, then gently pat dry on paper towels before using. In small bowl, beat egg white only until it starts to foam. Spread a thick layer of the sugar on a piece of waxed paper. Using a small pastry brush, lightly coat both sides of each leaf, with egg white, working with one leaf at a time. Gently dip each side of the leaf in the sugar, lifting the waxed paper to help pour sugar over the entire surface of each leaf. Set the leaves aside to dry on waxed paper at room temperature, about 2 hours. Leaves will keep for up to a week if refrigerated on waxed paper in an airtight container. Use frosted leaves to decorate desserts or drinks, or enjoy them as a natural after-dinner mint. Yield 36 leaves

Basil Jelly

1 quart water
2 cups firmly packed finely chopped basil
1 (1 3/4 ounce) package powdered fruit pectin 3 drops green food coloring
5 cups sugar

In a large saucepan, bring water and basil to a boil. Remove from heat; cover and let stand 10 minutes. Strain and discard basil. Return 3 2/3 cups liquid to the pan. Stir in pectin and food coloring. Return to a rolling boil over high heat. Stir in sugar. Boil for 1 minute, stirring constantly. Remove from heat; skim off foam. Pour hot liquid into hot sterilized jars, leaving 1/4 inch headspace. Adjust caps. Process for 15 minutes in a boiling water bath.

Mint Jelly

1 1/2 cups packed fresh mint leaves and stems
2 tablespoons lemon juice
2 1/4 cups boiling water
1 drop green food color
3 1/2 cups white sugar
1/2 (6 fluid ounce) container liquid pectin

Rinse and sort the mint leaves, and place them in a large saucepan. bruise them well with a potato masher or the bottom of a jar or glass. Add the boiling water, and bring the mixture to a boil. Remove from heat, cover, and let stand for 10 minutes. Strain, and measure out 1 2/3 cups of the minted liquid.

Place 1 2/3 cups minted liquid into a saucepan. Stir in the lemon juice and food coloring. Mix in the sugar, and place the pan over high heat. Bring mixture to a boil, stirring constantly. Once the mixture is boiling, stir in the pectin. Boil the mixture for a full minute, stirring constantly. Remove from heat, and skim any foam off the top using a large metal spoon. Transfer the mixture to hot, sterile jars, and seal. Process half-pints 10 minutes, pints for 15 minutes, in a water bath canner.

Prepared Horseradish

8–10-inch long piece of horseradish root
2 Tbsp water
1 Tbsp white vinegar
Pinch salt

Dig up an 8–10-inch long tuber of horseradish. Remove the leaves from the root and rinse the dirt off of the root well. Using a vegetable peeler, peel the root and cut it up into 1 or 2 inch pieces. Put horseradish pieces into a food processor. Add the water, and process until well ground. At this point be careful. A ground up fresh horseradish is many times as potent as freshly chopped onions and can really irritate your eyes and nose. Strain out some of the water if the mixture is too juicy. Add the tablespoon of white vinegar and the pinch of salt to the mixture. Pulse to combine. Using a rubber spatula, carefully transfer the grated horseradish to a half-pint jar. It will keep for 3 to 4 weeks in the refrigerator.

You can double or triple the ingredients if you have a lot of horseradish to use up. If putting up in half-pint jars for future use, process for 10 minutes in a water bath canner.

APPENDIX

Farm-Scale Food Dehydrator Plan

This drying chamber holds eight 3 foot square trays, with 3 inch spacing between trays. A 20-inch box fan is positioned adjacent to the trays. The fan draws heated air from the upper level and blows it across the trays. A shroud surrounds the fan to assist in directing the air across the trays. Air exhausts through the vent door or re-circulates back through the upper level. The vent door regulates the re-circulation of heated air. Heat is provided by a 1500 watt, thermostatically controlled space heater located in the upper level above the drying trays. The exterior wall is made of 1 / 2-inch plywood, the interior wall 1/2-inch drywall and the frame 2x2 and 2x4 inch lumber. The wall cavity is insulated with polystyrene insulation.

During the initial drying stage, the vent door is opened wide. After the removal of free surface water (about 15 minutes), the vent door is closed in stages to increase the circulation of heated air. The relative humidity in the cabinet is maintained below 30% and the dry bulb temperature below 160 F. If the ambient air is sufficiently dry, the dehydrator can be operated without additional heat input with the vent door wide open. In this design, the fan is placed in the lower level, adjacent to the drying trays to improve the uniformity of airflow across the trays, top to bottom and side to side.

Dale E. Kirk's home size dehydrator (USDA Home and Garden Bulletin 217) uses 9 75-watt light bulbs as heater for 8.5 square feet of drying area or 79 watts per square foot of drying area.

The I-Tech Farm-Scale Food Dehydrator describe here uses a 1500-watt heater on 72 square feet of drying tray area or 21 watts per square foot of drying area. The lower temperature output of the heater results in a longer drying time. It is partly compensated by the higher air exchange rate. If more heat is required, use suitable burner.

Airflow rate:

The 20-inch box fan used in the this Farm-Scale Dehydrator circulates 2100 cubic feet per minute in free space. If the tray resistance reduces the airflow by half, this equates to 58 volume of air exchange per minute.

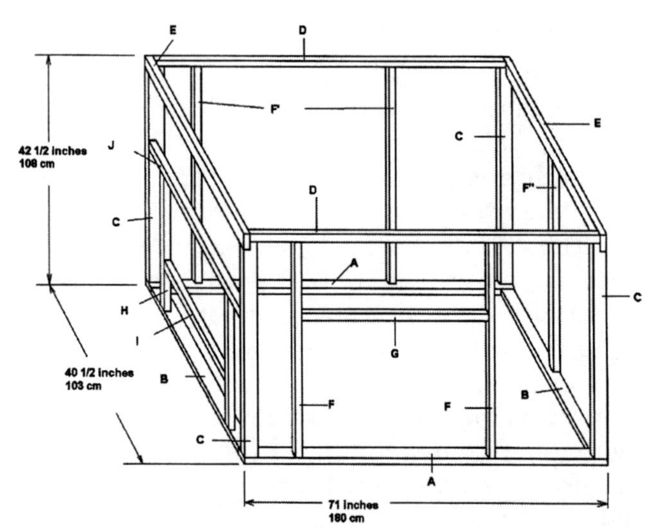

Figure 2. Dehydrator frame shown without 1/2 inch plywood outer shell, cavity insulation and '/2 inch interior drywall.

Not shown in Figure 2:

Half-inch plywood outer shell

Top and bottom, 2 pieces 72" x 41 '/2". Top piece is hinged to allow access from top

Front side and back side, 2 pieces 72" x 42 '/2" (43 1/2 h1 ll for top and bottom plywood). Cut out door

Ends, 2 pieces 40 '/2" x 42 '/2" ([41 1/2" 11'j x [43 1/211 —1"]). Cut out vent door and optional heater door

Frame: 2x2 miscellaneous lengths around the bottom for attaching drywall and optional gas heater opening.

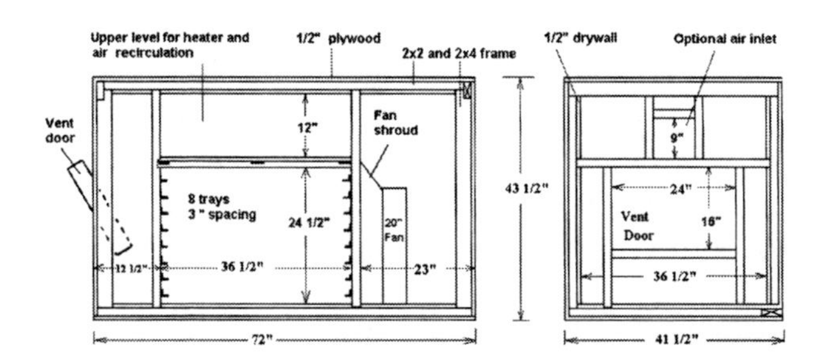

Side view End view
Figure 1. Food Dehydrator using '/2 inch plywood, 2x2 and 2x4 frame
and '/2 inch drywall.
Dehydrator components.
Frame 2x4 and 2x2 lumber, refer to Figure 2.

Bottom frame, 2x4, 2 pieces 71" (72" - 1" for plywood thickness)

Bottom frame, 2x4, 2 pieces 33 '/2" (41 '/2" - 1" for plywood thickness - 7" for 2x4)

End uprights, 2 x 4, 4 pieces 39 '/2" (43 '/2' - 1" plywood thickness, - 1 1/21 bottom 2x4, - 1 1/2 top 2x2), cut notch at top

Top frame, 2x2, 2 pieces 68" (72" - 1" for plywood thickness —3" for 2x4)

Top frame, 2x4, 2 pieces 40 '/2" (41 1/2" - 1" for ply wood thickness)

Opening for door, 2x2, 2 pieces 39 ½". Notch the 2 sidepieces to fit header. These sidepieces provide support for attaching the drywall, drywall divider separating upper chamber from lower drying area, and support for mounting drying tray bracket. F', 2 pieces, located in the back wall provide support for attaching drywall, drywall divider, and support for wood to mount drying tray bracket. F", 3 pieces (only one shown in Figure 2) located at end provide support for attaching drywall.

Opening for door, header, 2x2, 1 piece 37".

Vent door opening, side support, 2x2, 2 pieces, 28"

Vent door opening, bottom, 2x2, 1 piece, 24"

Vent door opening, top, 2x2, 1 piece, 37 ½"